NATURAL HISTORY of
COGNITION

NATURAL HISTORY *of* COGNITION

—— MIND OVER MATTER ——

CHUCK BAXTER

Copyright © 2020 by Chuck Baxter.

Library of Congress Control Number:		2020915697
ISBN:	Hardcover	978-1-6641-2396-0
	Softcover	978-1-6641-2395-3
	eBook	978-1-6641-2533-9

All rights reserved. No part of this book may be reproduced or transmitted in any form or by any means, electronic or mechanical, including photocopying, recording, or by any information storage and retrieval system, without permission in writing from the copyright owner.

Any people depicted in stock imagery provided by Getty Images are models, and such images are being used for illustrative purposes only.
Certain stock imagery © Getty Images.

Print information available on the last page.

Rev. date: 08/26/2020

To order additional copies of this book, contact:
Xlibris
844-714-8691
www.Xlibris.com
Orders@Xlibris.com
816670

CONTENTS

Foreword vii

Preface ix

Chapter 1   Cognition and Our Future: Why Would a Naturalist Study Cognition? 1

Chapter 2   Natural History of Cognition 18

Chapter 3   Is Consciousness a Property of Evolved Agency? 31

Chapter 4   How Do You Get Life from Physics? 49

Chapter 5   Mind-Body Duality & Causality 67

Chapter 6   Deep Learning and The Default Mode Network 83

Chapter 7   Life is a story: Poetics, Metaphor, Bayes, History of Models, and Reflections on the Ontology of Concepts. 107

Chapter 8   Virtues, Evolution and Life 129

Chapter 9   Religion for Century 21 145

Chapter 10   Behave Yourself: The Conscious Self as Mentor 164

Glossary 181

Acknowledgements 193

About The Cover 197

FOREWORD

As long-time friends of Chuck Baxter—some of our friendships go way back to the early 1960s—we wish to open this short Foreword with a statement of our deep appreciation for Chuck giving us the privilege to join his ambitious, often mind-bending, journey exploring the realm of consciousness. He has welcomed us warmly into his thought processes and given us the opportunity to read early drafts of the chapters. Through many hours of discussion and back-and-forth exchange of texts, he has offered us the chance to hone the profound and wide-ranging analyses he has developed over the course of many years. What a stimulating and inspiring mentor and friend he is!

The book represents a remarkable synthesis of ideas from fields ranging from philosophy to physics to cognitive neurobiology to evolutionary biology. It is in the context of evolution where Chuck's analysis truly shines—and confronts us with a novel perspective on what "drives" evolutionary change. Those of us who have enjoyed the many discussions of this book with Chuck are mostly biologists by training. However, not until we began our analysis of Chuck's text did we think so deeply about the pivotal role that cognition—as defined by Chuck in this book—could play in evolution. Rather than trek down the conventional path of evolutionary biology, which commonly uses DNA as its starting point, Chuck focuses on cognition and behavior, specifically the question of how experience leads living systems to perform better through "behaving better" in a fundamental and very broad way. Chuck's analysis is comprehensive and multi-tiered. The principles he develops apply at once to biochemical systems, food-seeking bacteria, and human societies. Life's capacity for learning from experience, which he discusses in the context of Bayes Theorem and contemporary analyses of several philosophers, including

Karl Friston, is what Chuck views as the essence of what evolution is all about. Chuck presents the case that whereas DNA, proteins, and other molecules may provide the chemical building blocks for evolution, the processes we define as "life" are shaped through adaptive changes in cognitive capacity and behavior that have been finely tuned by natural selection. After you have read this book, your view of evolution, and perhaps even life, will be radically transformed.

The level of erudition of this book is remarkable and it may challenge readers who lack familiarity with some of the concepts Chuck presents and integrates into his holistic analysis. But persevere, reader; there are great rewards in reading and, especially, pondering at some length the novel ideas presented in these chapters that follow.

We again thank Chuck for being such a wonderful friend and mentor over many, many years. He has helped us see the world differently and, in the process, helped to focus our thoughts on issues whose intrinsic fascination and mystery direct the mind into a realm that allows us to transcend, at least temporarily, the woes of the contemporary world. But this book is not in any sense 'escapist' reading; Chuck's motivation to write this volume was fueled in large measure by his belief that application of the rules underlying evolution can and should be seized upon by humanity to chart a more rational and humane course for the development of our societies. Evolution can teach us critically important lessons in virtue and morality. We have the power to change our minds and in so doing better our behavior.

So, reader, enjoy, learn from, and be better motivated for good behavior by the wisdom of this volume!!

Greg Baxter, Nancy Burnett, John Cooper, Mark Shelley, George Somero, Tierney Thys, James Watanabe

PREFACE

In 2006 I began a journey to discover what consciousness was and what it did. For 78 years I had presumed consciousness was at the basis of my decisions and perceptions. Then, I discovered that conscious awareness of a decision lagged behind the neural processing that produced it by ½ second or more. Also, instead of being as rich in content and detail as I had thought, my consciousness was limited to miniscule fragments of the neural processing that underlies decision making and perception. How could I run my life with only retrospective fragments of my complex dynamic world? In some way, however, it seemed my consciousness compelled and motivated my quest to discover the adaptive purpose of this enigmatic entity. I was not the first to embark on this venture, which has been going on for millennia. Moreover, I found that in the last few decades attempts to understand consciousness had become an exponentially popular journey. The landscape traversed by others was littered not only with evidence and parts of the story, which helped me, but also with claims of solutions, which often seemed unfounded. Thus, because these solutions did not fit my criteria for having reached a reassuring destination, I continued the quest. The terrain was difficult, filled with mazes and many routes that took me into strange intellectual territories where they spoke in foreign tongues.

As a naturalist, I came with different preparation than others on the quest, who often had strong backgrounds in cognitive neuroscience, philosophy, computer science, artificial intelligence, etc. However, in addition to the challenges inherent in my background, I found opportunities to use that background to discover and exploit little-explored perspectives. With a broad knowledge of general biology, I had often employed the comparative approach, which focuses on shared common features of different organisms.

This led me to the types of evolutionary and developmental analysis that considered similar processes in systems with different levels of complexity. Since consciousness was a feature of the brain involved in adaptive behavior, I elected to explore behavior in both complex and simpler systems. I asked, "What, in general, was the function of consciousness in influencing how life interacted with its environment to satisfy its needs?" One early waypoint was bacterial chemotaxis: the searching by the bacterium *Escherichia coli* for sugar in the water that bathed it. I found that *E. coli*'s general behavioral framework provides a simple model for evolved behavior. This perspective imparts common cognitive properties to all behavior when it is treated in the context of agency— i.e., behaving to attain a goal. My book argues that cognitive abilities are a characteristic of the evolution of life that is necessary for the agent—the organism—to interact with its environment to gain needed energy, resources, and a place to live. This interaction is evident at multiple levels of the hierarchy of biological organization, ranging from biochemical pathways to complex nervous systems present in networked, multi-agent systems.

In the following chapters, we will explore the details of life's evolved and hierarchically complex cognitive systems, eventually working our way up to humans with our individual lives as autobiographical memories. Through their conscious representations, our memories and attention interact with the ongoing programming of the paths producing the story. My consciousness put me on this journey when I became upset with its delay and miniscule size. I propose that consciousness functions as a director that influences the writing and editing of one's personal story. When I began this endeavor, I was perplexed by humans having unlimited capacity to produce problems that they then seem unable or unwilling to solve. We leave these problems as developing catastrophes for future generations to cope with. This is a problem in behavior, so I was prepared to examine the nearly 4 billion years of life and its behavioral solutions to problems in order to solve this quandary. Life has both worked to prevent as well as to recover from catastrophes. What can we learn from life's vast adaptive responses to environmental threats as it changed its behavior to cope? Life had to be rational and can provide us with heuristics from successful experiences in solving problems.

How a rational life can sometimes produce irrational outcomes led me to consider the importance of values, meaning, priorities, and emotion. This in turn led to the chapter on how ethics and morality evolved in biological systems. In human culture these also led to spirituality and religion. My chapter takes the perspective of a natural philosopher, which encompasses science, humans, and their cultural interactions with the natural world. It involves a feeling of reverence for Earth and the continuing story of life's adaptive behavior; it is a view based on an appreciation for life's creativity. Those tiny fragments of our active minds that appear in the consciousness provide each of us with an opportunity to script a life with a story that makes for a better world. We should all question if we are behaving well in scripting our lives because we must face reality with honesty. In the collective stories of our lives, we are in a race to separate authors of fiction from those who are dedicated to nonfiction, when seeking to solve complex problems.

CHAPTER ONE

Cognition and Our Future: Why Would a Naturalist Study Cognition?

"Advancing the minds of our youth with the growing science of the times. . .may ensure to our country the reputation, the safety and prosperity, and all the other blessings which experience proves to result from the cultivation and improvement of the general mind." Thomas Jefferson 1821

The mystery of consciousness.

It has been assumed there is deep philosophical understanding of the nature of consciousness, of what is real, if we have free will, the nature of justice, and the essence of life. Austrian-British philosopher Ludwig Wittgenstein (1889-1951) took the position that all facts relating to these issues are available to science and much of philosophy merely consists of disputes over word meaning. The scientific perspective replaced the philosophy of creation with mechanisms of evolution. Evolution of heritable variability in structure and function proceeds through competition with other systems within and between lineages. The correction of errors in adaptive behavior has helped ensure a lineage's future. Large and rapid changes in behavior can appear when natural selection repurposes existing structures for new purposes and shapes them to meet new evolutionary demands. Darwin had explained in principle that all current life may be traced back to an ancestral cell. Decades of research and sophistication in genetics, development, biochemistry, and molecular biology led to philosophers being assured by science that there was no need for **vitalism** to account for life processes. Science and philosophy have come to consider the advances gained through reductionist approaches to biology as removing the mystery as to how physics becomes biochemistry. However, we are still unable to do the same for the human brain. We can make all of the physiological, biochemical, and molecular observations on the operation of cells in the brain as have been done for other types of cells, but human introspection still strives to explain the mystery of how the constituents and activities of the central nervous system generate our consciousness. Conscious behavior in adult humans has a history both in its development and its evolution, so we will examine behavior from its very origins in an attempt to remove the mystery.

Behavior invented genes.

During most of my academic career, I followed the accepted paradigm of analyzing problems and gaining understanding through reducing one's focus to ever-simpler interactions of a system's simpler components. In my later years, I became more involved with complex interdisciplinary problems arising from multiple probabilistic causes that are also influenced strongly by their histories of timing and place of action. This late expansion in my career occurred when I became involved in trying to teach in a course in Holistic Biology. I recognized then the importance of analysis by synthesis that combines top down and bottom up approaches. Explanatory threads leading downward became webs of relationships leading upward. Similar insights had arisen in Robotics and Artificial Intelligence (AI) as they studied unsupervised learning in complex and variable environments. They created programs where perceptions were translated into adaptive actions. This is wholly analogous to the problems of life, which to survive must have internal models that can detect and adaptively respond to variability in its environment. Models begin as observations and inference of causes. The models are then tested iteratively by more observations and experiments to correct errors in their initial responses. This is formally treated in AI as Bayesian optimization, where models are treated as hypotheses judged against their performance in making correct predictions. I will treat **Bayesianism**[1] as a naturalist, leveraging off some recent—and often quite challenging-to-understand—literature that provides new conceptual frameworks for examining behavior and learning. For instance, **Karl Friston** developed his **free energy model** to describe how the brain works, where free energy is a more tractable stand-in for model error. He also proposes that similar functions apply in general to life's interaction with its environment. Life processes in the operation of the cell are treated by Friston as approximate Bayesian optimization of the inferential process of **generative models**. This broad point of view regards cognitive functions as mediating biochemical behavior as well as behavior of the organism as a whole. This relationship is a requisite for life and is manifested in the processes of homeostasis, resource acquisition, and replication. My thoughts here may seem contrarian or counterintuitive. *Most people maintain the perspective that genes invented behavior, whereas I believe the opposite to be true: behavior invented genes.*

In this book we thus recognize and document general principles of cognitive function that manifest in behavior at all levels of biological organization, from biochemistry to complex neural processes. Representation of a metabolite as a resource in the cell's internal environment by its biochemical model is equivalent to perception of objects in neural systems. We can reflect on the metaphysics of life. We experience behavior and introduce the general mystery of information as sense data and the 'feel' of consciousness. Few, however,

---

1 Many of the key terms and concepts used and developed in this book are highlighted in bold in the text and defined in the Glossary at the end of the book.

reflect on what the representation of the world 'feels like' in the biochemical model, but there must be a functional difference used as evidence for error correction, decisions, and control for the ongoing actions. What do differences in behavioral function 'feel' like to a biochemical pathway? Perhaps all the cognitive functions are present in the living cell minus the ability to talk about it. Of course, there is a vast chasm between a single cell and a human in how the complexity of the knowledge and information is perceived. The ability to talk places an agent in the realm of the philosopher, who translates unobservable functions into the mystery of the meaning of words. This book takes a natural history perspective on life's origin, consciousness, reality, sense-data as properties, free will, justice, knowledge and values; these all contribute to behavior in living systems. In summary, for nearly four billion years, life has focused on predicting how to behave and then behaving in response to input from a variable environment. It has evolved the ability to correct errors when its predictions have not been optimal.

Since this is a book about cognition, it deals with its product, which is behavior. *I synonymize cognition, behavior, life, and evolution as involving the same process, which is* **agency**. A **cognitive agent** is an autonomous system that acts to achieve its goals. The concept also overlaps into computer programs, artificial intelligence, and robotics. *We shall treat cognition as independent of platform and regard it simply as the ability to successfully model, in an interactive manner, another system.* This occurs because the agent can benefit from intervening in and exploiting other systems. Good behavior is achieved through monitoring performance and correcting errors in the agent's model. Behaviors can be chained together for more complex goals, but this requires networking for optimal interactive performance. **Error correction** is the key to excellence. As such, the detection and correction of errors is the function to be optimized. Errors are treated in four ways by models: 1) they can be ignored; 2) they can be discounted; 3) they can be corrected by moving or waiting for favorable conditions; and 4) they can be used as evidence to correct the model to better match the desired behavior.

To err is human—but so is error correction.

The book was motivated by the observation of societal malfunction in the face of our human-induced threats to our future. As a society, humanity has suffered through the consequences of delayed action on many anthropogenic problems. This avoidable suffering has caused me to reflect on societal rationality and our views on the future. What we accept as evidence, what constitutes an explanation, and how we understand causality must have some representation in cognitive processing. Rationality and cognition should lead to adaptive behavior when based on evidence, morality, and justice. How can we move society to recognize we are postponing the true long-term costs of our actions for the current short-term benefits to a few? Consequences of our environmental impacts are being discounted and true costs have been shifted to our children and coming generations. The function of rationality and justice involves the detection of errors in their early stages and their projection into scenarios that play out in true future consequences. There must be conciliation between the disparate views on these problems to maintain a functional civilization.

Shortly after 9/11, a former student and dear friend of mine, Rafe Sagarin, went to Washington D.C. to serve as an intern in the US Capitol as a Roger Revelle Fellow and watched as security safeguards proliferated. He observed many flaws in the largely uniform and predictable systems there and he let his mind reflect on ways these systems could be defeated and improved. Rafe, as an innovative biologist, recognized parallels with the natural world. He saw how for billions of years, organisms in their competitive survival mode had been confronted with security and defense requirements and been required to design innovations to deter attacks. He used his experiences in the tide pools and the knowledge of evolutionary biology and natural history to serve as models for antiterrorist strategies. His efforts evolved into organizing a published symposium, writing a popular book titled "*Learning from the Octopus*" and serving as an invited continuing lecturer on the topic at the Naval Postgraduate School in Monterey, California. His approach has caused me to reflect on the terrorizing effects of our environmental problems and our lack of effective action. When it comes to environmental issues, our media coverage documents failures in rationality, lack of error correction, and misaligned values which relate to special interests' interactions with our representatives in our government.

Biology can teach us how to behave.

Biological systems have succeeded by using experience to solve problems. Problems have been faced by living systems since life's origins, and perhaps we have something to learn from life's problem-solving strategies. Biological systems behave interactively with their environment as agents. In their networked interactions within individuals and in interactions with other individuals (agents), living organisms are social multi-agent systems that have mediated adaptive and rational behavior for billions of years. The complexity of using information in mediating rational decisions in the biochemistry and behavior of an individual is immense. This decision-making isn't just chemistry. It's adaptive regulation of hierarchically-arranged behaviors involving multi-agent communication to execute specialized behaviors as an integrated, functional whole. This book views this ability as living systems making predictions about their future, since we are always deciding what to do next. Detection of errors in our behavior and making decisions on values are cognitive functions. Since life has faced and solved environmental problems for nearly four billion years, our history has prepared us. Like all of life, we have the appropriate tools. We must commit to employ them. What can we learn from the history of life on earth to better understand and manage the current challenges to our success as a species?

I grew up believing that there is a way the world works, and it was my job to discover it. This is what produced my developing life story as my experiences became encoded into interactive memories in my consciously reviewable autobiographical life story. This story developed as my commonsense interpretation of the linkage between perceived information and its success in predicting interactions with the environment. This process is discussed by American philosopher Wilfrid Sellars (1912-1989) in *Philosophy and the Scientific Image of Man* where he discusses man's **manifest image** of the world. Sellars defines this image as the framework within which we ordinarily observe and explain our world as persons and things before the intrusion of the esoteric scientific image that may develop later. Our manifest image develops from observations and hypotheses about causes and meanings. These hypotheses are checked against repeated observations and experiments to correct errors, and appropriately modified through the shared observations and beliefs of others. Sellars contrasts this image with what the world looks like to us when it is dissected, expanded, and its actions are analyzed by the procedures of science. To illustrate this contrast, the manifest image of the chair you sit on is that it is strong and solid, while the detailed scientific view from modern physics is that it is 99.9999% empty space with some fermions and bosons bouncing around in fields of force.

Science as a way of knowing challenged the manifest image 500 years ago with our sun-centered solar system. For most of cultural history, myth has predominated as the means of explaining events whose causes eluded our direct perception. This all changed when the physical and later the biological world started being analyzed scientifically. Our manifest image of the world started to be replaced with models of new dimensions in language, scale, and time. These new models related to structure and process and required reformulation of our manifest image. During my academic years I recognized that viewing the world through such a realistic lens had adaptive implications and societal value. But the complexity of the world, combined with the complexity of science make describing the real world a Herculean task. This has led to the enigma of how we can ever solve problems in a completely rational manner when society always lags behind in understanding fundamental evidence and remains content with its manifest image. The manifest image is based on our conscious experience of the world, and while science can provide detailed causal perspectives on almost everything, the mind-body duality and physical basis of consciousness remain elusive.

The complexities of dealing with these questions about human minds motivated me to seek answers in the functional behavior and adaptive evolutionary origins of simpler systems. Most people have a view of living systems as structures based on genes; however, I have the belief that life must be understood as behavior. This derives from reflection on Nobel prize-winning physicist Erwin Schrödinger's definition of life as an open structured system that feeds on negentropy (negative entropy) and excretes positive entropy. Schrödinger stresses that we can recognize life by the fact that it continues behaving. This behavior is based on uniting heritable biochemical pathways with work cycles to participate in adaptive traits resulting in success and selection. *This claim specifies that life is behavior that is continually optimized by correction of errors in performance, to remain in a favorable relation to a complex and dynamic environment.* We have all lived our lives attributing causality to our experience with the world. Causality and its relationship with purpose are fundamental to explanations and the understanding we need to conduct our lives. I propose that exploring causality goes all the way back to the origin of life. Using causality provides the ability to intervene on the physics that we select and use to structure and act out our lives. Life tells physics what to do and how to do it to accomplish life's goals by creatively using physics' rules—but without breaking them.

This process led to the incredible story of the evolution and diversification of biological systems that has evolved over nearly four billion years and eventually produced the human species. Throughout this time, recognizing errors and refining corrective practices were critical for survival and success. Through the creation of hopeful futures at the individual level, this process produced the layers of diversity that structure the history of life on earth. This error-correction was largely through intergenerational evolutionary change. As members of a group that has arisen after three billion years of adaptive refinement of biochemical behavior, we have become large active creatures with the innovation of a nervous system. A nervous system enhances real time adjustment of behavior by improved sensing and interaction with the environment. We might view this advance as generating "online" adaptable behavioral programs. Here, of course, we speak of the group known as animals (Metazoa), which have proliferated and diversified in sea, land, and air in an incredible variety of forms and lifestyles and include us.

Adaptive behavior at the societal level—continuing the 4 billion year-old process.

Humans' success as a species has resulted in the emergence of our culture and civilizations, as well as a restructuring of ecosystems and some physical systems in our world. Our society has also created enclaves of minds that share alternative views of how the world should work and how to define serious errors in our behavior. Our society has sometimes selected leaders who neglect to update their beliefs based on evidence. Regrettably, we at times have placed such individuals in primary decision-making roles dealing with our most complex issues. Some leaders have even proclaimed they are without errors in decision-making, even in the face of documented deficiencies in performance. To continue as a species, we must use evidence from our history to make predictions and plan for a secure future. Societal impacts on our life support systems bring us ever closer to the time when we must choose to be architects of a secure future or wind up as victims of our own folly.

Behavior, according to control theory, is only accomplished by construction of a model whose parameters correspond to parameters of the system with which it interacts. Optimization of models through error correction, according to Bayesian principles, requires realistic probabilities of alternative outcomes. To achieve this, evidence must be openly selected and applied without ideological bias. My interest in Bayesianism derives from cognitive studies that identified Bayesian processes to be consistent with the creation of adaptive knowledge in our brains. *Bayesian brains continually monitor input and combine that input with knowledge from experience to creatively adapt existing behavioral models and make predictions about what should be done next.* This is the iterative process by which we lead our lives moment by moment. What gets especially complex is when large problems unfold, and we lack the experience and knowledge to quickly respond in an adaptive way. Understanding this challenge to our decision making becomes ever more critical in a very complex and dynamic world, where predictive skill of intelligence is required to decide the future direction of our planet.

Bayesian processes change models in an adaptive way in a complex dynamic environment. In this sense, "adaptive" refers to a value determination. This renders a Bayesian optimization of shared values as the critical first step that must be reconciled in operation of legislative and judicial systems. Through either ignorance or self-interest, these systems get evermore out of sync with what is of value to citizens. Citizens also continually fall further behind in acquiring and employing the information necessary to make rational decisions on so many important emerging issues. This is especially true for issues involving special knowledge, probabilistic interactions, and those issues playing out in high dimensions of interaction, space, and time. The amount and complexity of the information bearing on such decisions makes it possible for campaigns of misinformation to be effective in waging war on informed opinion. Problems are enhanced by actions of well-funded special interest groups and their ideological alignments with organized political groups. Four billion years of rational decisions have led the way to our civilization, which now has emerging problems placing severe time constraints on appropriate societal action. Decisions that postpone adaptive responses will result in future catastrophic scenarios. This delay raises the moral issue of our accepting responsibility to plan for the future of others who will be impacted by our decisions; we need to focus on intergenerational justice.

Knowing how things work requires the detection of causality, and when things are not working right, involves correcting errors in the causal chain. People grow up with a belief in causality and the feeling that there are reasons why things happen. Detection of causality is the basis of life's evolution and adaptation to the environment. In the human species and our emergent culture, causality has been the guiding force, not only for the individual, but for what should determine group decisions. For complex events, causality is not blatantly obvious but rather probabilistic. When multiple causality occurs, its participation will be probabilistic in both type of causation and degree of effect. How such probabilistic analysis is carried out is of critical importance in both assessing the use of current information as evidence and in choosing how to make decisions for a more deterministic future. Shortfalls in these complex probabilistic analyses have been the causes of all environmental problems and have produced a history of failure to anticipate and correct errors before high cost problems occur.

We err in two distinct ways: Type 1 and Type 2 errors in probabilistic analysis.

These shortfalls in use of probabilistic analysis in planning for the future are intimately involved with the treatment of Type 1 and Type 2 errors. The key challenge in these predictive efforts is in deciding when accumulating evidence tells you there is a developing problem. Science stakes its reputation on gathering credible evidence to support claims and beliefs. Its members thus abhor **Type1 errors**, which are false positives claiming that the evidence supports some belief when, in fact, such support is lacking. Publication generally seeks 95% confidence levels to avoid claims that may not be true. Sometimes this criterion for acceptance of evidence raises the risk of **Type 2 errors**, which is failure to conclude that an effect or relationship exists when, in fact, it really does.

Those who have economic or ideological commitments to their beliefs commonly are only open to evidence that either supports their belief or can be used against opposing views. In this instance, there can be little concern for the validity of the evidence in question. Such individuals make their living claiming that, for example, environmentally based objections to their actions and beliefs are based on Type 1 errors. Biased and fabricated evidence on complex issues requiring specialized knowledge lies beyond the comprehension and detection of non-specialists and opens the playing field for deception and confusion. For environmental issues, those who deny that harm is caused by some activity have used this tilted playing field to continually defer action until more "scientifically sound" evidence is available; in the meantime, the problem becomes ever worse. However, even when the evidence becomes overwhelming, they remain committed to resistance since they can't admit to errors in judgment. Their reputations and liabilities have become ideologically locked into denial of causation and the resultant harmful consequences of the activities they continue to defend.

Degree of potential harm prompts concerned scientists to invoke the **precautionary principle** and explore new and effective ways to inform the public and politicians as evidence becomes suggestive. In 2001, the war on terror was unleashed on a massive Type 1 error concerning Iraq's involvement in 9/11 and that country's possession of weapons of mass destruction. There have been about 260 deaths per year of US civilians since 2001 due to terrorism, with 80% of these in the 2001 event. The CDC informs us that 480 000 of our citizens die each year from tobacco use; these deaths result from the tobacco industry's fight against the use of evidence to solve a critical health problem. Both of these issues have been heavily involved with media, special interest groups, governance, ideologies, and emotional decision-making. The war on terror has accumulated a cost of over 2.4 trillion dollars, which has been discounted by adding it to our national debt. Can we tell if we are winning the war on terror or only making it worse?

Many of us feel that there has been an abuse of error types and risk assessment and cost benefit analysis. These suggested approaches to rationality do not play well in the political arena. Fear is a political tool and its emotional value speaks to amplify and inflate risk. The conservative position regarding acts of terror was exemplified by Dick Cheney's position that we faced threats of low probability and high impact. He proclaimed that if there was a 1% chance of Pakistan providing nuclear material to Al Qaeda, then this act should be treated as certainty. However, with somewhere between 95% and 99.999% certainty of anthropogenic climate change placing humanity at great risk, only about 30% of conservatives accept the evidence. Physics has allowed scientists to predict the fossil fuel contribution to climate change since the 1890s, which only leaves a question of how quickly it will develop and how severe its effects will be. Measurements and modeling have accurately projected scenarios of just how greenhouse gas emissions will play out in observed climate effects. These effects were established long ago, but climate change still continues as an ideological conflict in politics, which is responsive to corporate influences. Bets on the future are best made using real evidence to assess probabilities and this is especially true when betting lives. Rationality in cognitive action becomes more critical and also more complex in our society; it is time for us to exploit the tools and knowledge provided by our history.

Why choose ignorance?

> "The culture and civilization of consciousness has celebrated huge triumphs, but it also creates huge problems. The more power consciousness has over existence, the greater the problem of its paucity of information becomes. Civilization fills people with a sense of otherness and contradiction, which leads to the same kind of insanity we find in dictators surrounded by yes men." Tor Nørretranders. *The User Illusion*

Epigraphs will provoke different thoughts in different readers, so it might be wise to explain why the above text was selected to head this section. I chose this quote because it comprises two sentences in the last chapter of a book where I had already read 70 pages on aspects of information theory and, due to uncertainty about where the author was heading, I cut to the "punchline" of the book to search for motivation to continue reading. **"The User Illusion"** by Tor Nørretranders contrasts the display on a computer's screen with the structure and processing that goes on behind the screen and uses this contrast as a metaphor for human cognitive function. Thus, our consciousness is the selected screen display whereas the structure and functioning of the unconscious neural processes that create the experience are the deeper and much more complex computational activities of the subconscious brain. This metaphor or analogy implies consciousness is a limited retrospect of a complex underlying process of which it was unaware and opens the question of what purpose consciousness might serve, perhaps other than our entertainment. The last sentence of the epigraph is a synopsis of the xenophobia that plagues civilizations where people seek refuge from rapid change. The amount and speed of change in society has been amplified by an explosion of information and its complexity. Media sources proliferate through 24 hour news cycles, smart phones, and the Internet. An individual's input from these sources is estimated to have increased five-fold since 1986. Unfortunately, this explosion of data has not always resulted in the general public being more well-informed on emerging issues. Groups of people align with others who share their selected views and values and reject the perspectives of others. It was concern over why the conservative mind rejects overwhelming evidence of climate change and other looming societal issues that launched me into cognitive studies; *The User Illusion* opened a lead to a cognitive perspective. This initial step in my journey led to a decade of search for the origin, function, and ontology of consciousness and an attempt to understand the role, problems, and opportunities of the conscious mind. What can such analyses tell us about how information can become knowledge with meaning? In our evolutionary origins, we all had equal access to the information from the environment we occupied and those who incorporated information into better behavior were the ones who succeeded and persisted. That brief scenario summarizes nearly four billion years of life's story. With hominids' expanding social culture, groups or individuals made discoveries about some key aspects of the informational world and reached higher levels of performance than others. It was found that these advances could be taught through interpersonal information exchange. This was done increasingly within special groups, and knowledge became a powerful tool to enhance the status of privileged subsets of a population. As civilization flourished and proceeded to create sophistication in language and writing, it would make valuable information an increasingly powerful commodity. Books were painstakingly copied, and libraries of culturally selected and stored information were only available to a select few. In ~1440 Johannes Gutenberg democratized access to information with his breakthrough invention of a printing press with movable type elements. Where previously all books had to be hand-copied, now they could be mass-produced and made available to ever-wider audiences.

This progress was followed by restrictions of content and access by authorities who wished to control access to information, and thereby structure and control what was decreed as the best knowledge for the public. Information is infective, and as scientific breakthroughs began to cast doubt on the validity of authority, the supremacy of tradition and dogma were challenged. Print, literacy, and discovery interacted with culture, science, and philosophy and led to the challenge to both church and state for intellectual freedom. The founders of the United States of America were products of the Enlightenment and transplanted the rebellion against authoritarian beliefs to American shores. These efforts all emerged from the use of science as a model for discovery, where evidence and reason provided the guidance to truth. Instead of accepting tradition and the dogma of authority, an individual could use information as a tool to structure his or her own choices in what to believe. Our founding fathers felt that education was crucial to sustaining a democracy. Jefferson informed us on January 6, 1818: *"If a nation expects to be ignorant and free, in a state of civilization, it expects what never was and never could be."* Almost 200 years later, on February 24, 2016, candidate Donald

Trump proclaimed, after his victory in the Nevada primary election: *"I love the poorly educated!"*

Despite an increasing need for thoughtful reflection about the growing challenges our activities have created, at this period of time in the US, it appears that promoting ignorance is the goal of many politicians, corporations, media outlets, and socially aligned groups. Hundreds of books, thousands of editorials, and a plethora of opinions have already been expressed relating to this topic, but I enter the discussion through the path of a cognitive Bayesian who is interested in the relations between our consciousness, which is what we think we know, and our unconscious, which is the repository of all of our experience stored as knowledge. I developed this perspective when presented with two revelations. The bandwidth of information processing in our consciousness is of the order of 40 bits per second, whereas the unconscious cranks away at 100s of millions of bits per second. For both our perceptions and our willed decisions, there is a half-second or more delay from their beginnings as neural activity to their appearance as conscious awareness. The unconscious brain creates the thoughts and, through appropriate timing adjustments, makes them appear as though they shape our beliefs. This is a huge issue to get one's mind around, and the last ten years of delving into this phenomenon have convinced me that *our consciousness acts as an editor in its interactions with our experiences to create and replay the autobiographical story we use to navigate our way through a very complex world*. Our consciousness thus participates in the role of writing and editing 'our' story of the world and the role of 'self' as a character. The limited bandwidth of our consciousness places a premium on what is selected and by whom, to enter and participate in 'writing one's story'. How do we choose what feels right and how do we cast ourselves as a character in our story? These choices in turn determine how you want to write the story and who you want to be. So, choosing and processing of information becomes critical to how we view our lives. This activity relies on interactions between one's genetic inheritance (one's "nature") and developmental history (one's "nurture")—and both with Bayesian inference and rationality in setting criteria for truth and values in beliefs. This elicited my original question — Why choose ignorance?

What should be the role of the editor in the conscious mind in making decisions? It would seem that correcting errors would be a primary function. In order to correct errors, they must be discerned as a difference between your current belief and reality. This presents the challenge of using all the evidence available to create and update the probability of your belief. The motivation to be correct in a belief is often thwarted by the **Dunning/Kruger effect** where the **less** you know about something that you think you know, the more **sure** you are about your belief. Thus, the caveat: Don't have confidence in uninformed opinions. This is reinforced by the commonsense *illusion of the multitudes*, where illusory faith in their traditional knowledge supports errors on more complex topics. To quote Einstein, *"Common sense is the collection of prejudices acquired by the age of eighteen."* When your opinions are aligned with and supported by faith in inappropriate authorities, unreasonably high probabilities are placed on beliefs that are not based on solid evidence. Thoughtful comparisons and the careful sorting of expert opinion are essential when dealing with complex and technical subject matters.

Unwarranted confidence in one's beliefs exists at all socioeconomic and education levels. This state of faulty belief is what demands constant care and feeding of the unconscious, if you want to be right—or just less wrong. The unconscious uses Bayesian principles of rationality when motivated by curiosity to search for all the relevant evidence. Thus, desire for truth, questioning, curiosity, and motivation to learn are most important editorial qualities of the functional conscious mind.

Today, with searchable information widely available on the Internet, the choice of path becomes a personal decision. The physicist Richard Feynman tells us why science is the only human endeavor that improves over time, with the definition, *"Science is the belief in the ignorance of experts."* This is the critical attribute of Bayesianism: the emphasis that informed beliefs must be probabilistic and depend on all the evidence, not just what you choose as evidence. A continued questioning in a complex dynamic world is required since time's arrow makes new evidence replace old evidence. Ronald Reagan, who served as the 40th US president from 1981-1989, proposed the concept of Trickle Down Economics, where if you gave the rich more money, they would invest it and the profits would flow over to benefit all. That concept has been discredited repeatedly, yet with over thirty years of experience of tax cuts for the rich, the most successful middleclass in the history of the world has declined precipitously. When active individual questioning of unjust but reassuring dogma is avoided, the growth of happy ignorance extracts its toll on all of society and threatens the structure of democracy. For democracy to flourish, it requires justice, truth according to Bayesian analysis, and adaptation to a changing world.

Why discerning fact from fiction is so important.

Much is made of fake news, where people on both sides of an argument use facts selectively and fabricate different narratives with varying adherence to the truth. This creates people who have alternative beliefs about the same circumstances. My thesis is that, if you use your mind, then fake news is simply background noise obscuring the information you require to estimate the probabilities that you need to test hypotheses that can become beliefs. For nearly four billion years, life has operated and thrived by the selection of appropriate evidence that prompts rational positions. Life has also suffered severe consequences using inappropriate evidence. Life evolved in an environment with a vast amount of potential information. Restricting information's availability can be used to benefit the goals of the gatekeeper. In our society there is no unlawful restriction of access and the abundance and diversity of sources has never been greater. What supports the spread of fake news is really the public's choice of the source of their evidentiary news. If they don't like it and label it fake when, in fact, it is true, this makes them complicit in their own deception. Each person is responsible for creation of his or her own truth, and morality and values will play a role in this process. I will never understand why conservatives unconditionally love all embryos but have little regard for them when they become children. The cost of war, national debt, health care, social programs, education, environment, climate change, xenobiotics, racism, women's rights, and hope for a better future are critical issues for children. Conservative ideologies based on history and resisting change on these issues will result in children's future problems being exacerbated by conservative views. Conservative commentator William Buckley (1925 - 2008) expressed the philosophy: *"A conservative is someone who stands athwart history, yelling Stop"* in 1955, the same year he founded the *National Review*. The pace and magnitude of change has greatly accelerated, as conservatives rally to take us back to practices which treated critical issues in simpler times. Buckley was an intellectual conservative, however: *"Conservatives pride themselves on resisting change, which is as it should be. But intelligent deference to tradition and stability can evolve into intellectual sloth and moral fanaticism, as when conservatives simply decline to look up from dogma because the effort to raise their heads and reconsider is too great."* National Review, June 2004.

Further Reading and References.

Friston, K. (2017) The mathematics of mind-time. *Aeon* 18 May 2017 https://aeon.co/essays/consciousness-is-not-a-thing-but-a-process-of-inference

Nørretranders, T. (1998). *The User Illusion: Cutting Consciousness Down to Size*. Viking, New York.

Sagarin, R. (2012). *Learning from the Octopus: How Secrets from Nature Can Help Us Fight Terrorist Attacks, Natural Disasters, and Disease*. Basic Books, New York.

CHAPTER TWO

Natural History of Cognition

Minds and bodies: Reconciling our subjective and objective worlds.

There has been a long history of humans struggling over the problem of reconciling their subjective and objective worlds. The subjective world comprises private, experiential, ineffable feelings and thoughts that cannot be seen, measured, or experienced in any direct way by others. It is thus limited to largely inadequate verbal reporting, which others translate into attempts to correlate with their own subjective world. A student of cognition would replace *don't* with *can't* in the reply; *"You don't know how I feel."* However, we all work to become attuned to the thoughts and feelings of others, and how they influence behavior. This is the **mind-body problem**, which has produced studies and theories on the mind in the search of an explanation of consciousness or even its very existence. This problem has created feuding camps resulting in a plethora of papers and books and no general agreement.

The search for neural correlates of consciousness has employed a variety of experimental approaches and produced important insights, yet many questions remain. For example, there remains disagreement over whether consciousness occurs in human infants or other non-human animals. These are some of the core problems that have long intrigued me—and, eventually, led me to write this book.

Some autobiographical context: Why I set out on this quest.

I thought that I would work my way through life without writing a book, especially a book in the esoteric field of cognition and the mind-body problem. This is a field where everyone thinks they are an expert (since they can introspect), yet no one is really an expert (because of the complexity and diversity of perspectives on what the brain does and how it does it). A Google search on books on consciousness came up with over 50 million results, which I didn't inspect for individual titles. I have around 200 books, along with six or seven thousand downloaded research papers. Approaches to this topic range from the whole universe being conscious to consciousness as a quantum event.

I have spent my life reading books rather than writing them, perhaps for the same reason that I spend more time listening rather than speaking. At some point, I perceived that I learned more listening, especially if I was appropriately selective about whom I listened to. These habit patterns acquired early in my life were not conscious thoughtful decisions, but, perhaps, the type of thing that would be casually explained by referring to me as introverted, while not realizing that was naming rather than explaining. In my fourth-grade class, the teacher set up a reading challenge over the summer vacation. Our local library kept track of books checked out, and oral book reports were given of their content. That I had "wasted" my summer was evident when I won with 42 books; second place was 7 books, and the other twenty or so students lagged well behind second place. I also had the tendency as a child of spending more time communing with nature rather than indulging in sports - I was myopic, and better at examining bugs than seeing baseballs in flight. My early childhood interactions with nature and books probably had significant impact in programming the later journeys my mind would lead me on.

After a lifetime of teaching and trying to interest students in the wonders of nature, the Western Society of Naturalists certified me on the millennium as the Naturalist of the Year. Naturalists wander through nature observing and asking questions. Questions about what is out there, how does it work, and where does it come from? When naturalists get philosophical, they also ask "What does it all mean?" The only way I can justify writing another book on cognition and consciousness is that I bring a natural history perspective to the pursuit. I examine the topic in light of evolution and the role cognition and consciousness play in adaptive relations of life to its environment.

My academic career has been based on acquiring depth in a broad diversity of biological disciplines. As I matured, I became aware that these diverse disciplines shared many overlapping principles. To interest students in neurobiology, I would tell them, "Evolution designed brains to use history to predict the future." On reflection, I recognized that evolution did this for *everything*. In ecology, I began to recognize it was events in history that were reflected in the present structure of communities as they adaptively changed.

On reviewing physiology, I could also imagine behavioral ecology as an extension of homeostasis. Thus, the adaptive behavior of species interacting with their environment creates the structure of their communities, and these interactions translate into ecosystems. The basic rule of nervous systems is 'neurons that fire together wire together', and these interactions form networks that mediate perception and behavior. Through memory, these subroutines are structured by life experiences into more complex networks of interactions that govern the life of the organism. Evolution tells us that adaptations that work together are bundled together to assure a functional organism. Thus, biochemistry that works together wires together: Metabolic pathways that perform diverse functions coordinate the activities of their enzymes to ensure that the chemical fluxes needed for life occur in the appropriate directions and at the correct rates. At higher levels of biological organization, it is seen that all life depends through time on interaction with its ecosystem.

For me, this perspective on interaction across biological systems began to give depth of meaning to naturalist and author John Muir's (1838-1914) statement, *"When we try to pick out anything by itself, we find it hitched to everything else in the Universe."*, and geneticist and evolutionary biologist Theodosius Dobzhansky's (1900-1975) insightful remark, *"Nothing in biology makes sense except in the light of evolution."* However, despite these philosophical perspectives that gave such meaning to my analyses of biology, in order to prepare students for other courses and their futures, I focused on teaching the latest reductive paradigms of the expanding frontiers of science. This would all change after my retirement.

I taught at Stanford University for more than three decades and spent the last 20 years before retirement at the delightful setting of Hopkins Marine Station - Stanford's marine laboratory situated on the shore of Monterey Bay on the central California coast. I officially retired in 1993; however, my wife, Susan, was still working in the Stanford library, and I continued associations and close friendships with colleagues at the marine lab. I was a frequent visitor to Hopkins while going through several post-retirement ventures. Hopkins has a tradition of TGIFs to celebrate making it through another week. Those that choose to participate gather by the shore, with fermented malt beverage in hand, conversing and enjoying the natural beauty. One Friday afternoon in the spring of 2003, I was participating with a very small group that included my friend Bill Gilly, a professor at the station. Gilly had recently begun research on the Humboldt squid in the Gulf of California. As part of his Baja experience, he read *Log from the Sea of Cortez* by John Steinbeck and Ed Ricketts - recounting their 1940 cruise to explore the rich and diverse fauna along the shores of the Gulf. The book treats the biology, the people, and the magical setting of this inland sea bordered by a lush desert, as it recounts the story of the journey interlaced with holistic philosophy. Many outlandish topics arose at those TGIFs, so when Gilly proposed rerunning their cruise, I responded with great enthusiasm—figuring it would never happen. Gilly is persistent, however, and over the next year he assembled a skipper with a vessel, funding, supplies, permits, and a crew including me.

The trip for me was mind-altering in creating new directions of thought. Perhaps being on a vessel in blue water with blue skies overhead bears out the proposed interplay of that color with creativity. During the last leg of the voyage, Gilly met Susan Shillinglaw, a Steinbeck scholar, and their discussions led to the creation of a course for Stanford undergraduates in holistic biology. Instead of the standard reductionist-analytic approach to understand a major theme i.e. taking one perspective and pursuing it in depth of explanation, holism broadened the view, diversified the disciplines, expanded the dimensions, and sought understanding through synthesis. My participation in the course did all of these things to my mind and ultimately led to this book.

The course, Holistic Biology, was full time for an entire academic quarter; we spent 5 weeks at the marine station and 5 weeks traveling in Baja. The Hopkins section, with lectures, reading, discussions, projects, reports, lab and field work, along with social gatherings, began a holistic shift in thinking that students could then incorporate into their own projects carried out in Baja. I played the Ed Ricketts role of naturalist, ecologist, and philosopher which involved me in extended discussions of student interests. In the 2006 course, we had two students, Roddy Lindsay and Andrew Shaw, who were interested in cognition. I had read periodically in the cognitive literature and held the casual belief that whereas consciousness went along with neural activity, it was neurons that did the work and made the decisions. I was a mental materialist and lived through my conscious experience and recognized the role of perception in my interaction with my environment. However, I could not account for the actual machinery of my feelings of experience, or what it was they represented, or what they did, since they must be the product of my neural activity. I knew enough neurobiology and philosophy of cognition to participate in discussions with the two students; they had participated in Stanford's Symbolic Systems approach that incorporated philosophy, neurobiology, psychology, computer studies, and artificial intelligence in exploring the cognitive realm[2]. In Baja, Roddy loaned me Roger Penrose's book *The Emperor's New Mind* and this book introduced me, at the right time and the right way, to the work of **Benjamin Libet**.

Who's running the show? The Libet delay phenomenon, implicit learning, and the user illusion.

Libet's studies involved comparing the timing of the beginning of neural activity recorded from the exposed brain of conscious patients to both sensory perceptions and willed actions. Participants were asked to report the exact moment of their conscious awareness for both perception and a free-willed movement. His results showed a half-second lag occurred between participant reports of conscious awareness and the start of neural activity that mediated the event. This is known as the **Libet delay phenomenon**. Individuals make mental adjustments to compensate for this lag in order to be in sync with their perceived and willed world. For example: in baseball, a batter who tries to hit a 100 mile per hour pitch provides a detailed report of his perceptions and decisions during the 0.4 second of the flight of the ball from the pitcher's hand to home plate. Libet, however, informs the batter that his perceptions, actions, and decisions preceded his conscious awareness. In other words, the batter was giving a retrospective account from his short-term memories. Yet, the batter felt consciously aware of the pitcher's arm movement, the hand as it released the ball, the path of the ball, and even the rotation of the seams. Though he may have used all this information in deciding his swing, the processing and use of this information resided in his unconscious. This should not surprise those of us who, after driving a car on a familiar route while immersed in other thoughts, had no conscious awareness of our control of the vehicle and decisions in navigation.

These studies by Libet have huge implications for the concept of **free will**, which plays an important role in society, philosophy, and law. Before

---

2   Symbolic Systems Approach explores the relationship between natural and artificial systems that represent, process, and act on information.

reading Libet's study, I had regarded consciousness as simultaneous with the neural activity underlying perception and decision making. This delay of consciousness totally perplexed me. I began to consider the evolved function and mechanism of this cognitive adaptation. I lay awake all night in a motel in Santa Rosalia pondering the implications of a retrospective consciousness. I began to focus on an editorial role for selecting, motivating, and changing direction in the memories forming our autobiographical life story. I could only ponder and not research this topic in Baja. On my return to Internet access, I began to search for reactions to the Libet delay phenomenon and found it to be controversial, with authorities speaking out on both sides.

My perspectives were again confounded when my searches hit a paper by the British educationalist Dylan Wiliam in *Pedagogy, Culture and Society* that included a section on the bandwidth difference between the conscious and the unconscious states. For only sensory input, there is the potential for 11 million bits per second, and the ongoing background activity in the brain well exceeds this. In stark contrast to this number, the bandwidth of the conscious state is measured at only between 16 and 70 bits per second. I was astounded to discover that my consciousness knew so little about what my brain was doing. Also, if it were only these tiny fragments of our neural activity that entered consciousness, how could it seem so rich—and what selects the right stuff so it could play an editorial role? The batter, in the half- second of the pitched ball's flight, would have had several million bits processed by his brain, an amount of information that his consciousness could not have coped with, even if it was getting the information. Our brain is a beehive of activity, with only little fragments of the processing running through our conscious awareness.

In my reading, I was amazed to discover the explosion in the abundance and diversity of books and papers dealing with consciousness from the perspectives of physiology, psychology, philosophy, and artificial intelligence. During my academic career, it had been rather disreputable to even discuss consciousness, since many scholars regarded it as lacking in scientific credibility. This set my agenda for a search for understanding of brains, behavior, and the adaptive role of consciousness. If the mind had this complex time lag between subconscious initiation of a process and our conscious awareness of the event, as well as the capacity to selectively "feed" only a minute fraction of subconscious activity into conscious awareness and was consciously unaware of this, then there would appear to be critical evolved, adaptive roles of the consciousness that had yet to be discovered.

My reading of Wiliam's paper was intriguing since his interest was in **implicit learning** involving the unconscious, and I had developed interests in that direction from my participation in Holistic Biology. Implicit learning involves the adaptive shaping of behavior and understanding without awareness and reportability of what is learned; thus, it is intuitive in nature. Wiliam's paper had a reference to Danish science writer Tor Nørretranders (b. 1955) book, **The User Illusion**, in which the author draws an analogy between our conscious awareness (the "movie screen" in our brain), and the screen of a computer. In both, there is the display of a tiny fragment of the massive hidden information processing that creates what we consider the product of the machinery; and, in both cases, what we see is selected by the processes of the machine. It is also of interest that we use what we see on the screen to influence what the computer will do for us in the future. Early in my reading I became somewhat discouraged working my way through his introductory chapters on information theory and began to browse the final chapters to see if I really wanted to take the journey through the entire book. His take home message was that he was really approaching the emerging crisis of our mind's abilities for coping with the complexity of problems that society had created. This had been my focus for a number of frustrating decades, and I was struck by his words: *"The culture and civilization of consciousness has celebrated huge triumphs, but it also creates huge problems. The more power consciousness has over existence, the greater the problem of its paucity of information becomes. Civilization fills people with a sense of otherness and contradiction, which leads to the same kind of insanity we find in dictators surrounded by yes men."*

This view compelled me to try to understand a complex suite of relationships that could explain the following: (i) how information in the environment is selectively collected and made available for processing; (ii) what happens during unconscious processing; (iii) how this processing creates knowledge, meaning, and value and governs what emerges in the conscious, and (iv) how knowledge of this process can be used to facilitate rationality in both individual and societal decisions. Sometime earlier, I recognized the shortfalls in the general public's understanding of the science of emerging societal problems. I concluded that emotion, faith-based beliefs, and the excuse for ignorance called *common sense* were more important than evidence in these decisions. How will individuals and society respond to issues that are complex, multifaceted, and dynamic, without an understanding of a rational approach to solutions before catastrophes take the lead role? Since I was eighty years old at the time and retired, I could pursue this in the tradition of a naturalist motivated by a sense of wonder and explore and discover what cognition was all about.

A natural history approach to cognition.

The Oxford dictionary defines **cognition** as *"the mental action or process of acquiring knowledge and understanding through thought, experience, and the senses."* Explanations of our mental lives began as reflections on our beliefs, desires, and intentional behavior, and on the motivation to understand our interactions with others. In the folk psychology of society, terms and concepts appeared that became embedded in culture; these terms formed the basis of more rigorous work in the fields of philosophy, psychology, social behavior, neurobiology, and many other areas of interest that related to perception, knowledge, thought, memory, experience, consciousness, feelings, meanings and our behavior. This is only a partial list of concepts that are mental functions or properties and it confronts us with the yin and yang of cognition. What is the process and how does it fit the function?

Our quest for understanding these many issues related to consciousness creates problems rationalizing the subjective nature of the mental realm with science's causal physical explanations. This has led some into the **dualism** of **Rene Descartes** (1596-1650), who created a nonphysical and non-spatial substance for the mind as separate from the physics of the body. In examining beliefs and functions of the human mind, we should always be willing to confront the question "What is real?" It seems to us that the only reality we know for sure is the world and our thoughts as revealed in our consciousness. This information is used to create the stories we tell ourselves and others about what is real and what these stories mean.

To solve a problem requires an ontology to define what we are talking about, in order to both communicate information and relate to existing fields of study. Consciousness is obviously an important part of the ontology we are considering, so a clear definition is warranted. I define **consciousness** as an agent's engagement with the world in which the observation of a challenge or opportunity leads to the use of **active inference** in its resolution. Thus, consciousness is the ability of an agent to observe, monitor, and interact with its perceived world, and alter behavior in an adaptive manner, to allow improved interactions with the world. However, is consciousness itself real? And does it truly interact with behavior, or is it epiphenomenal and simply a reflection of the real underlying processes of physics and chemistry in the brain? Positions on this question remain diverse and controversial after 2500 years of reflection. While there may be conflicting views, science works to continually correct errors and to conciliate hypotheses that arise from different perspectives. Cognition studies are massively diverse and specialized; they range from philosophy, through all the human and non-human animal studies, to robotics, artificial intelligence and even quantum mechanics. Since we are the species of interest, are conscious, and are the only ones that can introspect and verbally report, cognitive studies have heavily focused on humans. It is suggested that, including the activities of our brains, humans may well be the most complex entities in at least our solar system. Some feel this complexity may pose insurmountable barriers of structure and function to an understanding of the mind in terms of physical causality.

The existence and diversity of our biological world was an equally complex problem when observations during the voyage of a certain naturalist provided the framework for evolution, which then enabled a deep and rational approach to the structure and function of life. As Charles Darwin (1809-1882) informed us, *"It is interesting to contemplate a tangled bank, clothed with many plants of many kinds, with birds singing on the bushes, with various insects flitting about, and with worms crawling through the damp earth, and to reflect that these elaborately constructed forms, so different from each other, and dependent upon each other in so complex a manner, have all been produced by laws acting around us."* Since cognition emerged in evolution, using the approach of natural history may provide insight into what it is and what it does.

At the millennial meeting of American Society of Naturalists, Peter Grant defined a naturalist as one who asked questions of nature and answers them by every means possible. A question is a problem to be resolved and requires agreement on defining what you are talking about. The subjective nature of many mental concepts presents problems in communication. Cognition deals with the acquisition and use of knowledge, which is the basis of behavior, and is treated in systems ranging from biochemistry to human social behavior to robotics. For a naturalist, this provides the opportunity to employ the **comparative approach** to examine systems ranging from simple to complex, which may occur on different platforms, e.g., in widely different species. From the study of systems common to diverse species, unifying principles about these systems can often be gleaned. Below, we will see how using the comparative approach in studies of cognition can assist us to develop broadly applicable principles that apply across a remarkably wide range of biological phenomena.

Comparative analysis of cognition: responses to hunger—from bacteria to humans.

A fundamental behavior of living systems is the acquisition of food or resources from the environment, and we begin our comparative analysis of cognition with a focus on feeding behavior. Below are two short stories concerning individuals who got hungry and went on a search for food.

'One morning, while on a trip to a meeting to talk on *E coli* chemotaxis, I woke in my hotel room, dressed, and went for a walk. I turned down a side street where some businesses were beginning to open, detected the aroma of donuts being cooked and felt the urge to indulge. Sampling the strength of the airborne signal, I zeroed in on a small shop where dozens of doughnuts were piled on a tray. I selected one still warm and began the search for a cup of coffee to be able to dunk before eating.'

'A bacterium casually exploring its environment felt its energy supply dwindling, and struck out to detect a sugar source, powering its movement with counterclockwise beating flagella. On detecting a noxious fatty acid, it reversed flagellar rotation, reoriented, and began on a more hopeful course. When the bacterium picked up the signal of sugar, it began swimming, comparing the detected scent of glucose from the last second to the previous three seconds, and, finding it higher, it stayed the course. Through a succession of runs and reorients, it worked its way up the gradient to the mother lode and tanked up on glucose to fuel its metabolism.'

These are both stories of an individual motivated to find a food source and their subsequent behavior to achieve a goal. The first involves a human, introspecting their autobiographical memory and recounting selected behavioral events of perhaps an hour out of a fifty-year life. Humans are very complex entities, composed of 37 trillion cells; that hour of behavior would involve the interactive processes of billions of neurons and trillions of synapses. The bacterium story may have been told by that same person at the meeting based on his or her laboratory observations of the behavior of a single *E coli* cell of one micron length, a cell that lacks a nervous system and has a life cycle of perhaps less than an hour. However, genetics and epigenetics produce structures and pathways that enable it to swim thirty body lengths in a second and make behavioral decisions about whether things are getting better or worse.

Typically, cognitive studies are directed at the structure and processes in the adult human brain and are designed to understand how nerve cells act to monitor our beliefs, desires, and intentions in performing behavior. A long cultural and scientific history introduced human concepts such as purpose, values, free will, the feel of subjective experience, introspection, thoughts, and both rational and emotional decision making into cognition. These properties have been presumed by many to exclude consciousness and psychological properties from the vast majority of behavior. As a naturalist interested in diversity, the evolution of life, and in models of behavior in computers and robotics, I choose to examine the broad spectrum of behavior to achieve understanding of its functional fundamentals. A general theory of the association of described subjective mental constructs with underlying behavioral functions can provide objectivity to the mental concepts and help in the path to understanding.

Nearly four billion years of evolution precede the separation of the human lineage around 5.8 million years ago. Other mammals differ from us by about 8% genetically (DNA sequence), and chimpanzees by perhaps less than 2%. A variety of animals are used as systems to study behavior, both in nature and at physiological and psychological levels in the laboratory. To tell the story of cognition, we must consider the diverse and complex behavior required to support the lives of species evolving and adapting in complex and dynamic ecosystems.

Cognition and agency: intake and processing of information and generating a 'point of view' needed for action.

In his provocative way, Cognitive scientist and philosopher Daniel Dennett (b. 1942) in his 1992 paper, *Time and the Observer*, reflects on consciousness as the action of an observer monitoring information from the environment in space and time, along with the experience of perception and behavior. *"Wherever there is a conscious mind, there is a point of view. A conscious mind is an observer, who takes in the information that is available at a particular (roughly) continuous sequence of times and places in the universe. A mind is thus a locus of subjectivity, a thing it is like something to be. What it is like to be that thing is partly determined by what is available to be observed or experienced along the trajectory through space-time of that moving point of view...."* This certainly fits well with the story of the gentleman seeking a doughnut. What if we paraphrase Dennett's mind as observer, by replacing mind with cognition, since it fits the Oxford dictionary definition of cognition given above: *"the mental action or process of acquiring knowledge and understanding through thought, experience, and the senses."* Then, since the bacterial cell has food seeking behavior, let's lower the bar to examine cognition in the behavior of a living cell, and come out with an *E coli* point of view.

'Wherever there is a living cell, there is a point of view. Living systems are observers required to use the information that is available at times and places in their environment to regulate biochemical behavior. Life thus becomes a locus of subjectivity, a thing it is like something to be. What it is like to be that thing is determined by what is observed or experienced along the trajectory through space-time of that moving point of view.'

Before you ridicule this broadened use of "cognition," reflect on the cell's behavior in homeostasis and its responses to challenges and opportunities in interaction with its dynamic environment. Dennett points out that a functional definition of consciousness is the experience of the observer and the subjectivity is the observer's representation of the environment it interacts with. We will make the case for this as a generalized cognitive approach to all life's behavior. Although this does not generate acceptance of anything like our subjective cognitive experience occurring in a living cell, it certainly joins together the function of observing environmental information in time and space and using these observations in adaptive behavior.

Thoughts such as these, along with discussions of whether thermostats are conscious, suggested to me that the early evolution of living systems may provide perspective on the evolution of cognition. In our broad approach to the understanding of cognition, we have equated it with the processes that implement behavior. Philosophy treats behavior as **agency**. A general definition of agency is the capacity of any living being to act in any given environment. Agency is considered as purposeful, goal-directed activity, which implies that an agent exerts a kind of direct control or guidance over its own behavior. Thus, in our search for the evolutionary origins of cognition, which precedes the origin of higher nervous systems, the journey takes us back to the emergence of agency in biochemistry. This holistic cognitive perspective will help us in treating the deep question of how evolution built the agency of living systems from physics.

Further reading and references.

Dennett, D., and M. Kinsbourne (1992). Time and the Observer: The where and when of consciousness in the brain. *Behavioral and Brain Sciences* 15: 183-201.

Libet, B. (1985). Unconscious cerebral initiative and the role of conscious will in voluntary action. *Behavioral and Brain Sciences.* 8: 529–566.

Penrose, R. (2002). *The Emperor's New Mind.* Oxford University Press, New York.

Wiliam, D. (2006). The half-second delay: what follows? *Pedagogy, Culture and Society.* 14: 71-81.

CHAPTER THREE

Is Consciousness a Property of Evolved Agency?

What distinguishes "life" from "physics?"

In 1944, Erwin Schrödinger wrote a small book from the perspective of a physicist titled, *What Is Life?* The book had its origin in a series of public lectures focused on the theme *"How can the events in space and time which take place within the spatial boundary of a living organism be accounted for by physics and chemistry?"* I read it in 1952, just after changing my major from engineering to biology and remembered one concept that I would recall throughout life: Life is a highly structured open system whose survival and maintenance is fed by negative entropy. In turn, life excretes positive entropy. Sixty years later, I read *Reinventing the Sacred* by Stuart Kauffman, who defined the separation of life from physics through the evolution of **agency**. He characterized organic evolution as uniting heritable biochemical pathways with work cycles and proposed that molecular autonomous agents provide a minimal definition of life. Living systems are thus molecular autonomous agents that adaptively structure biochemistry to convert negative entropy into the maintenance and replication of themselves.

My later rereading of Schrödinger was highlighted by his observation that the way you could tell if something represented "life" was if the entity in question continued behaving. The quest of these autonomous biochemical agents—living systems—is to seek out and acquire energy and resources from the environment. This perspective on life provides a conceptual platform to treat adaptive heritable biochemistry as the beginning of behavior. This conceptual platform can take us beyond simple biochemistry to the analysis of the human mind. New approaches to how the mind works consider it to be a networked part of the body, embedded in its working environment, and functionally interactive with external resources through adaptive behavior. Thus, we can regress nearly 4 billion years and see cognition emerging with life and its first biochemical structures and processes.

The **belief-desire-intention (BDI) model**, developed by the American philosopher Michael Bratman (b. 1945), emerged from folk psychology as a perspective for treating others as practical reasoning agents. From introspection, we come to think that belief and desire determine our intention, which, in turn, controls our actions. Human agency is the capacity of an actor to use a BDI model to predict that perceiving something they desire causes their actions. This has been a useful model and the concept has been employed in contexts ranging from biochemical pathways to computer programs and artificial intelligence. Agency in living systems is fundamentally based on biochemically based behavioral programs, optimized through error correction, that link information and action with resources. Multi-agent systems are a set of cooperative agents acting for a goal either within an individual or between individual agents. They develop value functions that enable development of a prioritized selection of input for subsequent action. Behavior is monitored for performance and error correction in a dynamic and competitive world. Implicit in this perspective is the realization that behavior is selected and thus becomes the basis of evolution and learning.

Regarded from the perspective of a naturalist, this conceptual viewpoint translates into the discipline of behavioral ecology, where, through perception, decisions, and actions, a species fits into and adapts to its ecological niche. Energy and materials must be acquired from a complex and dynamic world through the actions of adaptive behavioral programs. This conceptual framework also applies to biochemical pathways, whose programs must continually adapt in order to regulate the flux of intracellular energy and materials. Thus, physiological homeostasis and an organism's life cycle are also seen as instances of networked multiagency of structured behavior that leads to adaptive resource and energy utilization. Having designated agency and associated behavior as cognitive biochemistry, how can we reconcile this view with the psychological attributes we associate with our behavior as conscious agents?

According to the American psychologist Julian Jaynes (1920-1997), author of *The Origin of Consciousness in the Breakdown of the Bicameral*

Mind, consciousness has emerged as a philosophical concept in recorded history only since the Homeric period (1200 – 800 BCE). Jaynes proposes that before this period people were not self-reflective on the role they played in their lives. The voices of gods directed them through another part of their mind with guidance on complex decisions. As societies, communication and language became more complex, self-awareness and an internal narrative appeared as our conscious control. Since consciousness is rooted in self-reports of thoughts and feelings, the question arises as to how language, culture, and their interactions have influenced our concept of consciousness.

Many in the field of cognition disdain a definition of consciousness and direct readers to formulate their own definition through introspection, without cautioning them about influences of our history of folk psychology. What was there before we began talking about it? Could our concept be one of the stories we accept and tell that has grown within our culture, as in the cases of religion and the immortal soul? If consciousness evolved as a trait and there is pre-language consciousness, then Australian philosopher **David Chalmers'** (b. 1966) **hard problem** concerning how the machinery of the brain (its neurons and synapses) can produce consciousness (discussed in Chapter Six) and the enigmatic and long-running **mind-body debate** should be solvable. The major problem is getting rid of a conceptual homunculus as an assumed mechanism that can view, feel, and make decisions for us, and ceasing to assume that our experiential world resides in this unanalyzed control module. Where and how are our world entities perceived? Psychological aspects of behavior should be regressed down the chain of elements that mediate the behavior, and comparative analyses of simpler systems may reveal their emergence on more tractable platforms than our brains.

What do we mean by "explanation?"

To understand something requires an **explanation**; classically, scientific explanations involve hierarchical **reductionism** to basic physics. Thus, in science we seek explanations by taking a complex system and reducing it to a sequential chain of causal relationships that provide a stepwise set of already rationalized and accepted lawful relations of action. In the words of Nobel laureate theoretical physicist Steven Weinberg, *"All the explanatory arrows point downward, from societies, to people, to organs, to cells, to biochemistry, to chemistry, and ultimately to physics."* Another Nobel physicist, Murray

Gell-Mann, has defined a natural law as a compressed description of the regularities of phenomena. If we follow Weinberg's arrows down in a biological system, we find them mediating actions and interactions that are unpredicted, were it a system determined by the rules of physics.

In dealing with cognition and consciousness and developing explanations of psychological concepts in general, investigators focus on neural systems, predominantly those of humans. Traditionally, the treatment of consciousness has taken the perspective of sophisticated, traditionally educated adult humans. This educational experience provides the background for their introspections on the basic nature and structure of their perceived consciousness. However, they generally fail to consider consciousness from the perspective of its historical and developmental origins, and treat only the ultimate product, the adult conscious experience. The Dutch philosopher Baruch Spinoza (1632-1677), at the beginning of the Enlightenment, criticized philosophers for this type of approach in treating complex issues and suggested starting with the fundamental properties of nature and projecting them into production of the final product. Psychological attributes of genetically determined biomolecular pathways that show behavior are usually ignored or treated with derision in terms of their utility in contributing to understanding of behavior. American philosopher Thomas Nagel's (b. 1937) contribution to thinking about consciousness provided us with the fundamental dictum that in a conscious system *'there is something it is like to perceive and behave.'* This statement provides the perspective for comparative and evolutionary subjectivity and the potential for a science of consciousness. A natural history background in biology directs one to take complex systems back to their origins in development and evolution and search for general similarities between systems. Since the question of interest is how the brain determines behavior and 'what it is like' to behave, we should take behavior as the system to analyze for psychological properties. **Behavior** is generally considered to be the act of an agent doing something for its own purpose.

We must model the world to interact with it effectively.

The world is a very complex system, one for which we must create a **model** if we are to operate successfully in reaching our goals. Physicists will tell us that the world is fundamentally a quantum universe of probabilistic wave functions and elementary particles in motion, which we can only detect with instruments making measurements and explain with theories. Our sensed world of structure and interactions is based on an incredibly diverse, numerous, and dynamically complex set of assembled aggregates of these elementary particles. We refer to these as the "real world" of classical physics. We, humans, are constructed at that level, and our continued existence depends on appropriate interactions with this world, which constitutes our environment. The real world of our environment does not exist in our brain or any sensing system; rather, biological systems construct models to infer it. Living systems are based on sets of these models, which constitute their perceptual world and that are shaped through interaction with the environment.

Emergence, agency, and cognitive systems.

Life is an **emergent** system that can not violate the laws of physics; however, new rules for the organization of matter and flows of energy have emerged from the process of evolution. It is important to recognize that while in physics the laws determine the system, in evolution the system—life—determines the laws. Our Universe expands, unfolds, and structures itself according to a set of rules established at its very beginnings and that apply everywhere throughout time. Evolution began the development of a parallel set of rules for the control of matter and energy for the purposes of living systems. This involved the emergent properties of knowledge, its replication, information on concepts and attributes, values, and the ability to exert control over structuring matter and the direction of energy flows. Rules established by evolution were flexible and could be altered to correct errors and thereby work better in allowing organisms to extract materials and energy from their environment. These properties are packaged into evolved biochemical systems and have resulted in the emergence of agency.

Agents have the ability to control actions in their system rather than the actions being initiated only by the laws of physics. All elements of the behavior depend on physics and chemistry, but where, when, why, and how behavior happens violates the Gell-Mann protocol of natural laws and is not predictable. Consider a mountain goat and a boulder starting from the same spot and descending a steep slope. Gravity, terrain, and Newton's laws of motion determine the boulder's descent. The goat chooses a destination and skillfully selects a sequence of small precarious foot placements and may even change its mind and go back up the slope. Evolution has produced systems that in the material sense comply with the laws of physics but have added properties that are not physical. Therefore, reductive physical explanations often must be complemented by addressing "why" questions for a full understanding of the biological phenomenon to be attained.

Knowledge, information, values, adaptation, and control of action are emergent properties of the machinery of evolution and they produce agency. Agent 007, James Bond, is an agent with a store of adaptive knowledge and the ability to monitor information from the environment and control his responses in beneficial ways by predicting the future. We would also describe him a cognitive system. A general property of a **cognitive system** is one that uses models of other systems to adaptively interact with those systems. These modeled systems can range from physical properties such as oxygen levels or temperature, to sources of nutrients or other valued resources, to other complex cognitive systems within or external to the agent. Human agents and mountain goats are a long way from the origins of agency a little less than four billion years ago.

The origins of life, behavior, and agency.

The origin of life through a chemical evolution was perhaps first addressed by Charles Darwin in a letter to the British botanist Joseph Hooker in 1871: *"conceive in some warm little pond with all sort of ammonia and phosphoric salts,—light, heat, electricity present, that a protein compound was chemically formed, ready to undergo still more complex changes."* Russian biochemist Alexander Oparin and British scientist J. B. S. Haldane published theoretical scenarios in the 1920s speculating on life's origin in conditions on the early earth. In 1952, chemists Stanley Miller and Harold Urey simulated a supposed primordial atmosphere and exposed it to electric discharge. Beneath this atmosphere was a 'warm little pond' of water in which a variety of organic molecules, including sugars and amino acids, were quickly produced. This investigation also stimulated research and a still growing literature on the origin of life. Theoretical biologist Stuart Kauffman, in *Investigations,* offers a definition of the first life as a minimal molecular autonomous replicating agent that can employ a thermodynamic work cycle. Other concepts specify the ability of a chemical system to undergo evolution, cooperation between systems, or specify the appearance of something like RNA. Currently, we consider the cell and its characteristics as defining life.

Among the plethora of proposals on how chemistry evolved into life, biochemist Nick Lane (1967-) replaces the warm little pond on the earth's surface with the deep-sea alkaline vents, which develop slowly and may persist for over 100 000 years. Detailed in his readable book, *The Vital Question*, these grow miles away from the ephemeral and dramatic black smoker vents in the deep-sea spreading zones. These vents provide the time, structure, materials, energy, catalysts, and a porous structure that both facilitates reactions and allows accumulation of organic products. The vent fluids are alkaline. The bathing ocean waters back then were very acidic. These conditions established a voltage gradient across the porous walls of the vents that supplied an energy source for chemical reactions enhanced by mineral catalysts. Lane's story begins with these environmental proton gradients powering production of the first amino acids and sugars. Their concentration and potential as catalysts could support the beginning of biochemistry and the formation of simple polymers and elements of metabolism. Lipids provide the opportunity for membrane production, allowing simple protocells to appear. Transmembrane differences in pH could establish proton gradients to drive organic synthesis. Acquisition of hereditary capacity in these protocells may have occurred with their acquisition of RNA. Lane synthesizes a lot of biochemical detective work in support of his belief that the structure and processes of the alkaline vent and its longevity provide conditions for the progressive accumulation of stages and variety in biochemical evolution. The vent environment is viewed as serving as the "incubator" for the first cell, satisfying the prediction that Darwin made in *On the Origin of Species* in 1859, *".... that probably all the organic beings which have ever lived on this earth have descended from some one primordial form, into which life was first breathed."*

While biochemical evolution could have been diverse and happening concurrently in many ways and places, evidence indicates that all living organisms came from a single ancestral lineage that gave rise to Bacteria and Archaea, the two groups that would massively diversify and be the sole occupants of our world for two billion years. This hypothesized cell is known by the acronym, LUCA, for last universal common ancestor. This was an extremely simple cell proposed to have this set of basic characters: 1) a bounding lipid membrane, across which there were numerous ion gradients; 2) a loop of DNA with the triplet code, 3) RNA for allowing transcription of genes, 4) ribosomes for protein synthesis, 5) L-isomers of 20 amino acids, 6) metabolic pathways including elements of the Krebs cycle, 7) ATP for energy, 8) perhaps 355 genes for proteins, 9) replication of cell constituents and cell division, and 10) energy from hydrogen oxidation. There are many alternative stories of the origin of life, but I am attracted to Lane's scenario since it deals with energy powering biochemical behavior. *Here I recognize biochemical pathways as the origins of agency and cognition.* The resolution of the detailed events in the origin of life lies in the future, but we know it happened somewhere between 4 and 3.6 billion years ago with the appearance of LUCA. The two single-celled lineages, Bacteria and Archaea, that LUCA spawned evolved with basic differences in composition of cell walls and cell membranes, ribosome structure, RNA polymerase, and glycolytic pathways, yet both reflected the common attributes of all living systems, including biochemical systems that supported the evolution of behavior and agency.

Rather than approach evolution as a story of how biological structures evolved, I prefer to focus on a behavioral story. **Behavior** is information in action controlled by knowledge-based agency. Behavior can range from its most basic forms involved in the control of biochemical reactions to overt familiar acts of movement. My thesis is that evolution resulted from competition between replicating systems for limiting resources and produced agency, which is behavior with goals and values. The agent system was the basic unit of life, which could evolve and chain together in symbiotic relationships to produce ever more complex systems. This recursive process is marked by the emergence of increasingly varied and complex traits that produce and enable agency. Despite this growth in complexity and variety of biological agents during evolution, a cognitive behavioral analysis of agency across time and scales of complexity reveals striking generality in the fundamental characteristics of agents and comparable functionality of what we regard as their psychological aspects.

Agency has a huge literature, one that is mostly related to humans and their voluntary actions or a person who performs a function for another. We require a more general and objective definition, one that allows us to examine the properties, structure, function, and evolution of agents at different levels of biological organization. Indeed, agency has even been proposed in non-biological systems. Thus, in artificial intelligence, an agent is considered as an autonomous entity, which observes and acts on its environment, directing this activity towards the system's goals.

Organismal behavior and artificial intelligence: How similar are they?

In living systems, the autonomous entity in question represents an adaptively interactive model in an agent designed by evolutionary and/or learning processes to implement behavior related to achieving a goal. Our objective in this analysis is to recognize cognitive functions in behaving systems that are open to analysis and to relate these functions in a general theory of behavior. Biochemical pathways analyzed as behavior can provide a physical basis for the understanding of function in psychology. Artificial intelligence and robotics employ designed behavior where functions are open to comparison. Sanz et al., in their 2007 article, *Principles for Consciousness in Integrated Cognitive Control*, examined consciousness in the context of artificial intelligence and provided a set of links between agency and psychology. In item 6 awareness incorporates a value function in perception. Keep these in mind as general provisional definitions as we examine the behavior of an organism without a nervous system.

> *"1] A system is cognitive if it exploits models of other systems in interaction with them*
>
> *2] Model Isomorphism – An embodied situated cognitive system is as good as its internalized models are.*
>
> *3] Maximal timely performance can only be achieved using predictive models.*
>
> *4] Generating action based on a unified model of task, environment and self is the way for performance maximization.*
>
> *5] Model Driven Perception – Perception is the continuous update of the integrated models used by the agent in a model-based cognitive control architecture by means of real time sensorial information.*
>
> *6] System Awareness – A system is aware if it is continuously perceiving and generating meaning from the continuously updated models.*

> *7] System Self-awareness/Consciousness – A system is conscious if it is continuously generating meanings from continuously updated self-models in a model-based cognitive control architecture."*

Agency began with individual molecules, and even though they became complex, their future was limited to being "just chemicals" (recall Nick Lane's alkaline vents). The next wonderfully creative step was an aggregation of interactive molecules into a private space bounded by a differentially permeable membrane (LUCA). Here, the knowledge-based agency could evolve and regulate the internal environment as well as incorporating agent-based mechanisms in the membrane for adaptive bidirectional traffic flow through the membrane. The abilities to move in nutrients, use them for energy and synthesis, move out wastes, and grow and reproduce set off a massive radiation of specialized lifestyles. Some even developed proton powered motors driving filaments and producing movement of the cell. As we saw in Chapter Two, this ultimately resulted, at the bacterial level, in motile behavior with ability of the individual to choose its direction of movement adaptively in a complex environment.

Bacterial chemotaxis: biochemical cognition driving adaptive behavior.

Chemotactic behavior was slowly assembled over eons of time by the Darwinian process of building on mutations that were beneficial to biochemical and structural adaptations in existing agency. Acquiring, crafting, and the coordinated chaining of the products of the 50 genes used by bacteria for this behavior are comprehensible in a species that, under ideal conditions, can have a generation time of twenty minutes. In less than nine hours, this could represent a quarter of a billion individuals for mutation and selection to work on. Darwin was comfortable with the reproductive potential of elephants to drive selection and evolution, but for elephants to equal *E. coli* would require over 1000 years. *E. coli* became a model system early in molecular biology and has a massive literature relating its genetics to its biochemical pathways.

> *"E. coli, a self-replicating object only a thousandth of a millimeter in size, can swim 35 diameters a second, taste simple*

chemicals in its environment, and decide whether life is getting better or worse." — Howard C. Berg (Professor of Molecular and Cellular Biology, Harvard University).

E. coli has approximately 4300 genes that code for a variety of proteins, both enzymatic and structural. About 50 of these genes participate in the production of the system that mediates the bacterium's locomotion and chemotactic behavior. Half of this group of genes are for proteins that construct their locomotory structures. These are structures with proton-driven reversible motors bearing long filaments fabricated as 8 or 10 flagella. Counterclockwise rotation allows coordination of the flagella for a straight run at a speed of 35 body lengths per second. Reversal of the motors results in a brief tumbling and reorientation in direction of movement. In an uninteresting environment— one in which the local chemistry and physics pose neither opportunities nor challenges—this flagellar activity produces a three-dimensional random walk with about one second runs and 0.1 second tumbles.

The other half of the 50 behavioral genes code for sensors and a communication pathway to control the motors for chemotaxis, which is movement toward chemical attractants or away from chemical repellents. In a run on their random walk, the bacteria may detect the presence of an attractant or a repellent and will extend their run time. At 1 micron they are too small to detect a concentration gradient over their body, but they have a memory that determines whether the concentration increases when movement is occurring in a particular direction. Studies indicate that a cell compares the concentration observed over the past one second with the concentration observed over the three previous seconds and responds to the difference. The sensors detect a variety of both beneficial and deleterious environmental conditions and nutrients. The attractants include some sugars, amino acids, and dipeptides, while repellents are some alcohols, glycerol, some amino acids, high CO_2, and high and low pH.

When moving in a favorable direction with respect to the gradient, *E. coli* prolong the run time, but for an unfavorable direction they reverse the motors and tumble, to hopefully reorient. This results in a somewhat erratic but net movement to place them in favorable circumstances. Since they swim in a curve even when moving in a favorable direction, they tumble approximately every 10 seconds. Though their path is erratic, their persistence results in occupying situations that are beneficial to their wellbeing. After a period spent in the favorable environs, they will habituate and start the random walk of short run-then-tumble behaviors and begin to explore for new opportunities.

The diagram below presents the key elements of the regulatory model used by *E. coli* in its chemotactic behavior. The labels refer to proteins or their complexes encoded by the chemotaxis and motility genes. The default rotation of the flagella is counterclockwise and behavioral control is mediated by the timing of the switch that produces a brief clockwise rotation,

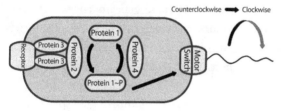

which leads to the resulting tumble. The switch is triggered by the level of phosphorylation of Protein 1. In an ongoing exploration, Protein 2 autophosphorylates and phosphorylates Protein 1 at levels that promote short runs and tumbles. Favorable environmental stimuli are sensed by the receptors and, acting through Protein 3, result in decreased phosphorylation of Protein 2. Protein 4 also acts to dephosphorylate Protein 1 to prolong counterclockwise rotation and stage the course in favorable conditions. On reaching a favorable site, a period will occur during which the cell exploits the chemical(s) it needs. The cell then will begin to explore for new stimuli and move from the previous attractant. Habituation during the period of resource exploitation is produced by mechanisms that both take the brakes off autophosphorylation of Protein 2 and inhibit Protein 4-mediated dephosphorylation of Protein 1.

The energy to drive the motors for chemotactic behavior is based on the voltage gradient maintained across the plasma membrane; this gradient drives a trans-membrane proton flow that powers the flagellar rotor. This elegant cellular machinery produces runs whose duration before tumbling and reorienting depends on environmental conditions that regulate pulses of phosphorylation of the motor's switch. This sensed information from the environment regulates phosphorylation levels of proteins, a molecular-level example of information flowing in a behavioral program. This is a simplified summary that both skips extensive details and processes yet to be unraveled. What this summary provides, however, is the outline of an *evolved biochemical pathway that determines complex and labile behavior of an individual bacterial cell.*

A similar behavior performed by a planarian, insect, shark, or mammal, as the product of a complex nervous system, would be described in cognitive neural-behavioral terms. Behavioral descriptors such as search, motivation, decision, adaptation, satiation, preference, avoidance, and memory are recognized as products of complex nervous systems, but what is the relevance of such terms to bacterial behavioral machinery? What is cognition in a bacterium? Let us consider bacterial chemotaxis as chained hierarchical multiagency in operation—an integrated chain of sensing and locomotory traits. *Our general definition of cognition is the use of a model of another system to interact with it. Here the other system is chemical gradients that exist in the classical physics of the environment. The model is the structured biochemical pathway selected by evolution that recognizes information from a valued resource.* This information regulates a function, here, locomotion, in an adaptive way. *The use of internal models that living systems construct to relate themselves to the world they sense is a principle that we can apply at all levels of agency.* This exercise equates biochemical models with neural models in a functional way and provides us with a deep philosophical background to search for the emergence of cognitive properties, including the origins of consciousness with the evolution of life.

"What is it like to be...?"—a central question in analysis of consciousness.

The philosopher **Thomas Nagel** believes that a being can be viewed as conscious if there is "something that it is like" to be that creature, i.e., some subjective way the world seems or appears from the creature's mental or experiential point of view. Famously, Nagel places the discussion of consciousness in a form we can all relate to by posing the question, *"What is it like to be a bat?"* This is more complex than simply imagining our mind in a bat's body, and Nagel felt it exemplified the intractability of the mind-body problem. As practicing folk psychologists, we feel we can imagine ourselves in the minds of others, but even the reality of this for how other humans perceive and feel is suspect. Nagel's statement refers to what it feels like to the bat, not how it would feel to us as bat mind readers. *From bats to earthworms to jellyfish and bacteria, we cannot conceptualize their subjective world, but if they are conscious, Nagel proclaims there is something it is like to be them.*

We have a lifetime of experience in what it is like to be us, but for millennia the relation of the subjective to the material world has posed a puzzle that remains a mystery. The first levels of consciousness are essentially forms of what Ned Block (b.1942), an American philosopher dealing with consciousness, calls **phenomenal consciousness**. This form of consciousness relates to the experience of "what it is like" to perceive sense data, which then are employed in models that guide behavior. These activities thus comprise the experience of behaving in concert with one's perceptions, thereby exerting control of actions in the world one experiences. Nagel claims that there must be something this "is like" for each organism. Each of us experiences "what it is like" to be our self and can only wonder "what it is like" in others.

Can artificial intelligence help us understand bacterial chemotaxis—and vice versa?

After the above description of bacterial chemotactic behavior, we are in a good position to reflect on evolved agency as a cognitive system. Comparing the wide array of disciplines contributing to the cognitive literature, I have found the perspective of those working in artificial intelligence and robotics to be helpful in clarifying questions, providing definitions, and creating instructive model systems. Let us return to Sanz et al., and use their seven principles for machine consciousness as a framework for comparing bacterial behavior to the cognitive properties of complex systems.

> "1] *A system is cognitive if it exploits models of other systems in its interaction with them."* The receptors are models of means for identifying characteristics of sensed environmental resources, including the quality of the chemical and physical environment, and gaining information on how these resources change with time. This information tells the cell how to modulate its swimming behavior. The memory-based gradient detection model provides information on the distribution of chemical resources in a variable environment and on whether to keep swimming or flop. The gradient concentration-based regulation of phosphorylation levels' control of the motor's switch is mediated by biochemical pathways, which act to make decisions on adaptive search and choice of environment
>
> "2] *Model Isomorphism- An embodied, situated cognitive system is as good as its internalized models are."* These embodied

models have been situated in environments with challenges and opportunities. Adaptive behavioral models have been crafted by over billions of years of evolution. They are continually refined by selection across trillions upon trillions of generations to do their job.

"3] *Maximal timely performance can only be achieved using predictive models.*" The detection and response functions are the end result of a hereditary sequence of striving for predictive success. The interactive set of models has been tuned by evolution to minimize error and optimize performance in sensing and prediction of what to do next in adaptive behavioral responses in a spatially variable environment.

"4] *Generating action based on a unified model of task, environment and self is the way for performance maximization.*" Internal states of the cell and sensing relevant environmental conditions regulate the pattern of movement of the cell to work to locate it in the best place it can find for its current state. *E. coli* prefers glucose, but when glucose runs low and lactose is available, the cell uses information of these events to switch preference and crank up a lactose-digesting pathway by lifting repression of the gene coding for the involved enzyme.

"5] *Model Driven Perception- Perception is the continuous update of the integrated models used by the agent in a model-based cognitive control architecture by means of real time sensorial information.*" The real-time sensing provides information on what to do. Where to go is the role of short-term memory on right or wrong movement in the gradient and what to do next. This constitutes real-time updating of the model-based control architecture for chemotaxis.

"6] *System Awareness- A system is aware if it is continuously perceiving and generating meaning from the continuously updated models.*" The real-time sensing and short-term memory of relevant environmental conditions and gradients have value functions that exert control over the bacterium's interacting biochemical pathways. Since the individual cell responds continuously and adaptively in order to efficiently acquire resources as it moves through a complex environment, its cognitive control architecture thus exhibits **System Awareness**.

"7] System Self-awareness/Consciousness- A system is conscious if it is continuously generating meanings from continuously updated self-models in a model-based cognitive control architecture." The cognitive control architecture of chemotactic behavior generates meaningful behavior for the cell system. It does so by integrating the ongoing activity of all the linked biochemical pathways of the multi-agent individual. The chemotactic behavior is not stereotyped; rather, behavior will switch with changing conditions in the external or internal environment of the cell. All are controlled by the activity and interaction of biochemical pathways. If the cell does not generate appropriately adaptive meaning for the purposes of its integrated control architecture in competition with others, it loses to those who generated better meaning. **Meaning** is seen as the adaptive value to prioritize and predict appropriate behavior as conditions change. This activity requires communication within the network of models that link resources of its world to the homeostasis and cell cycle models that combine to form the self-model of the individual cell.

This analysis introduces the psychological concept of meaning to bacteria and computer programs. Meaning is necessary for the translation of the target symbols as information into the construction and use of the model's computational program. This program is for adaptive behavior that requires meanings for purpose and values. This is much discussed as the **Symbol Grounding Problem** of how words or symbols get meaning. Philosophically this is intrinsic to what meaning really is, but our interest is how mental states and consciousness have meaning that supports purpose and values in autonomous systems. Behavior requires a model that represents a program that manipulates the physics of the platform by learning rules to achieve a purpose. Our **generative model** for behavior selects the symbol system with a sensorimotor system that reliably connects its internal symbols to the external objects they refer to for interaction, thereby providing meaning for the behaving system. If meaning derives from process in behavioral models, this supports a general theory of autonomous adaptive behavior, a theory that provides linkage throughout evolution, learning, robotics, AI and reflections on artificial and extraterrestrial life.

A cognitive analysis of *E. coli*'s chemotaxis cannot give one an intuitive feeling for what perception or consciousness is or "feels like" in biochemical pathways. This sort of lack of understanding is an inevitable problem for construction of a general theory that treats biology as behaving systems with a purpose. The mystery of "what it is like" for other agents cannot be resolved. As we look at behavior of systems that mediate simpler behaviors that are more direct products of evolution and occur in groups separated by evolutionary distance, we are less inclined to posit awareness of subjective feelings. This is also the case with machine intelligence. We celebrate the diversity and complexity of morphologies produced in evolution and sometimes forget they were structured as behavioral adaptations.

Some thoughts on human ontogeny and development.

I am old enough to have heard a teacher say, "Ontogeny recapitulates phylogeny," before this notion was largely cast aside. Ontogeny can provide us with insights, however. If instead of an evolutionary perspective, we examine consciousness from a developmental approach, we again observe increasing behavioral complexity. Each human life begins as a single cell where behavioral events are at the same level as *E. coli*, but with more genes and different objectives. As development unfolds, the original cell divides many times, all the while becoming more behaviorally complex. Communication between groups of cells produces their movement and differentiation. This shapes the body plan with primordia for organ systems and morphological structures. These are brought into coordinated control through a circulatory system and a nervous system.

As the embryo begins repetitive and random motor behaviors, the mother perceives her baby kicking and turning; sonograms of the fetus demonstrate an ever greater variety of behaviors as development proceeds. Of course, it is after birth that we follow the development of the child into a rational communicating being with large scale neural development continuing for several decades. Comparatively, as a newborn ages we can align levels of behavior with that of other mammals. Thus, in preverbal infants we speculate on levels of consciousness. The mystery mentioned above occurs again: "What is it like" to be a child who can't verbalize?

Further Reading and References.

Figure based on Eisenbach M (2011) Bacterial chemotaxis. In: eLS. Wiley, Chichester. doi:10.1002/9780470015902.a0001251.pub3

Kauffman, S. (2008). *Reinventing the Sacred: A New View of Science, Reason, and Religion.* New York: Basic Books.

Sanz R., I. López, M. Rodríguez and C. Hernández (2007) *Neural Networks.* 20, 938-946.

Nagel, T. (1974). What Is It Like to Be a Bat? *The Philosophical Review* 83: 435-450.

Webre, D. J., P. M. Wolanin, and J. B. Stock (2003). Bacterial chemotaxis. *Current* Biology 13: R47-R49.

Lane, N. (2015). *The Vital Question: Energy, Evolution, and the Origins of Complex Life.* New York: W. W. Norton and Co.

Jaynes, J. (1976). *The Origin of Consciousness in the Breakdown of the Bicameral Mind.* New York: Houghton Mifflin.

CHAPTER FOUR

How Do You Get Life from Physics?

Adopting an evolutionary perspective provides a conceptual platform for understanding the origin of behavior and beginnings of cognition. Behavior, agency, and cognition are emergent properties from a set of ancient, heritable biochemical processes. These processes created physical systems possessing knowledge, goals, and the ability to work towards those goals. With this innovation, the way the world works was fundamentally changed as some matter behaved for its own destiny.

This viewpoint was my rationale in Chapter 3 for treating concepts of bacterial chemotaxis as functionally the same as terminology used to describe behavior mediated by nervous systems. In this chapter, we take a deeper look at the foundations of behavior, beginning with an examination of what I feel is life's most important equation.

Bayes' Theorem: life's most important equation.

In A Brief History of Time, Stephen Hawking recognized the negative correlation between the number of equations in a book and size of the book's readership. While he avoided equations, he did feel compelled to include Albert Einstein's theory of special relativity, $E = mc^2$. One of his reasons was that, along with the work of Newton, Einstein's relativity equation established the laws of physics as universal. This means that wherever you are you can understand the physical systems by the same tools we use on earth to analyze them and project their future states, with an accuracy that is a function of the data.

Physical systems don't care about and can't alter their futures, as they lack knowledge, purpose, information, meaning or goals. Those capacities emerged with life and are properties treated by what I regard as the most significant equation on earth. For life, the goal is to keep the system evolution has produced maintained and running whilst interacting with the variable aspects of a complex dynamic environment. Life achieves this through behavior that can act on knowledge with a purpose. Behavior can predict that if I can sense and act as I did in the past, it will be for the best. The accuracy of this predictive capacity is a function of how well the behavioral model works in a complex dynamic environment. Information and meaning help build the model and behavior corrects errors in the model's performance. Life is the process of using a history of successes and failures of performance in the past to predict the future through the continued correction of errors based on a conformance to principles of Bayes' theorem. Behavior is a probabilistic hypothesis about causality in a variable environment where differences between predictions and performance are used to increase predictability; this is formally treated by Bayes' theorem.

Following Hawking's lead, then, I will introduce only one equation, which I feel is to biology what Einstein's equation is to physics. It offers a conceptual process for shaping and correcting agency.

$$P(b|e) = P(e|b)P(b) / P(e)$$

Bayes' theorem computes the probability beliefs are true based on the evaluation of new evidence. This probability is the first term of the equation— $P(b|e)$. How should beliefs be based on this new evidence? The numerator is the probability of observing evidence given prior beliefs $P(e|b)$, which is multiplied by the probability of prior beliefs $P(b)$. The denominator is the probability of the new evidence $P(e)$. Bayes formally told us we should continue to use evidence for accuracy in beliefs.

Bayes' theorem addresses a key question: How much do we believe and how can we improve our beliefs? This questioning process is how we live our lives, and some of us are better than others in being realistic about the truth-value of our opinions and how evidence is used to make beliefs reflect reality. Life evolved through creating and testing the performance of heritable behavioral models in a complex and probabilistic world. Success was achieved by selecting models that performed best. Bayes' formula treats the interaction of knowledge about a given event with behavioral responses to it and judges whether performance got better or worse. Should I move, wait or change my behavioral model? If I want to behave rationally, I must learn from experience and correct errors. Evolution tells the story of expanding knowledge and competence based on selection of memories, goals, and successes. Bayes formula reflects the fundamental logic of life, evolution, behavior, science, rationality, and truth. Bayesian truth is characterized by a quote from environmental ethicist Bryan Norton in *"Searching for Sustainability"*, *"In complex dynamic systems the truth is dynamic and adaptive or else it is irrelevant."* Since the complexity of the real world makes causality and/or likelihoods of evidence difficult or impossible to compute directly, we must be both amazed and pleased that living systems conform so well to the predictions of Bayes Theorem. Thus, it has been employed in theories of models treating behavior and evolution.

Agency is based on Bayesian optimization of behavioral programs with an adaptive purpose. It links information and action with resources. Multi-agent systems develop functions that balance priorities to maintain the system (homeostasis) while monitoring the appropriateness of responses to a dynamic competitive world. Selection of genes depends on the performance of genetic regulatory programs playing roles in models of behavior. The behavior is focused on maintaining homeostasis of the individual based on its interactions with an environment that offers both resources and hazards. From the perspective of a naturalist, this translates into the discipline of behavioral ecology, where environmental resources of energy and materials must be acquired from the environment, which is also chosen for its suitability for the individual. You do not want to be a fish out of water; furthermore, only parts of your water world are appropriate. Success will depend on sets of models responding to opportunities and challenges during the individual's experiences. This conceptual framework also applies to maintenance of its constituent cells as well as its body as a whole. These can be regarded as internal environments for the hierarchical behavioral agent systems of the individual. Thus, both homeostasis and the life cycle of the organism are networked multiagency of behavior for adaptive resource and energy utilization. Or said another way: homeostasis and an organism's life cycle are the result of a multi-agent network of structured behaviors that adaptively use energy.

Karl Friston and his free energy principle.

The British neuroscientist Karl Friston (b. 1959) formalizes the emergent properties of living systems under his **free energy principle**, which involves approximate Bayesian optimization of the models for behavior. The free energy approach assumes biological systems must actively resist a natural tendency to disorder. This is essentially Schrodinger's view of life fighting entropic decay by feeding on negative entropy. In Friston's view, organisms as agents use internal models of their external environment to make predictions, which they then respond to through behavioral action. The concept is known in psychology as **belief-desire-intention (B-D-I)**. In psychology this is a model of human reasoning and has also led to software used in programming agency in Artificial Intelligence.

Organisms are intentional behavioral models with prior expectations of interactions with their environment. Their behavioral actions then become equivalent to hypothesis testing. Natural selection and/or learning improves their models by error reduction, thereby minimizing their free energy. Free energy is Friston's metric for a model's performance: low free energy denotes high performance. Each phenotype establishes its own niche by maximizing its model performance using evidence from its surrounding environment to maximize homeostasis and resource acquisition. Organisms thus develop prior expectations about interactions with their environment based on their history of success and failures. These expectations are termed **affordances** by American psychologists Eleanor and James Gibson in their theories of environmental psychology. Our history of interactions with our environment creates our functional perceptual world through evolution and learning. When organisms encounter surprises, they must either adjust their input by action or adjust their model to interact more appropriately. The free energy principle should resonate with those having a background in science where models are used in explanations. We can choose where to use our model and when it must be error-corrected to minimize its free energy.

Another look at E. coli and "biochemical decision making."

E. coli isn't able to do math or perceive an overall view of the chemical gradients it swims through. Consider these gradients as a hilly landscape where bacteria select or avoid peaks for adaptive reasons. The bacterium can't detect the shape of this landscape, but it carries out a sampling algorithm that enables it to assess the local landscape at four-second intervals as it moves. With the use of a biochemical model, this information tells the bacterium if its environment is probably getting better or whether it may be getting worse. These movements are the result of decision-making—i.e., deciding whether to continue swimming or reorient. These decisions enable it to climb or descend through the probability landscape. Formally representing and working with probability distributions of resource landscapes is not possible for E. coli, but repetitive sampling of the distribution is easy. Step by step, choosing a local sense of getting better, these behaviors lead the bacterium toward the solution. The algorithm that detects errors and decides what to do next was crafted over evolutionary time in a biochemical model that used evidence to correct errors.

How do biochemical pathways "change their mind," enabling a cell to be adaptive in a complex dynamic environment and to cope with the requirements for changes demanded by homeostasis and the cycle of the cell? We have presented E. coli's seeking for a source of glucose as a cognitive model of cell agency, but an agent must be flexible and adaptive in decision making. When glucose is in short supply and lactose is available, the cell, which previously disdained lactose as a substrate, can switch to use it for an energy source. This is accomplished through a chemical information pathway. When glucose is in short supply its phosphorylation when entering the cell declines and excess ATP is converted to the signaling molecule cAMP (cyclic adenosine monophosphate). cAMP along with a breakdown product of available lactose in the cell can lift the repression of the lac operon. This process enables activation of genes promoting both the uptake of lactose and its digestion. Thus, both signals for the lack of the preferred glucose and the presence of lactose are required for the decision for the change of biochemical behavior. This system of biochemical pathways can be described reductively as chemistry doing its thing, but in order to understand and explain its evolution and actions, we require the concepts of heritable knowledge and goals. These concepts are involved in rational behavior. Does a biochemical system that enables solutions of problems in adaptive behavior use functions that play similar roles in the expansion and diversification of behavior?

Daniel Dennett: "stances" to take in seeking understanding.

When confronted with a system whose behavior you wish to understand and predict, Daniel Dennett, in his book *The Intentional Stance*, proposes three strategies: the *physical stance*, the *design stance* and the *intentional stance*. The first involves physics and chemistry. Here, one isolates and analyzes systems and, from their constituents and properties, applies the laws of physics. This mode of analysis conforms to Nobel laureate physicist Stephen Weinberg's statement that, "*all the explanatory arrows point downward.*" Ideally in such a physics and chemistry based analysis, both understanding and prediction are excellent. The *design stance* falls into the domains of invention, engineering, art, and evolution. Here, something other than the laws of physics make the system work and cause it to do so for a purpose. Predictions are made from knowledge of goals achieved through the design of the system. The *intentional stance* is the domain of cognition, agency, and the mental processes, where information, knowledge, goals, values, and plans determine behavior of the system.

In Dennett's words: "*Here is how it works: first you decide to treat the object whose behavior is to be predicted as a rational agent; then you figure out what beliefs that agent ought to have, given its place in the world and its purpose. Then you figure out what desires it ought to have, on the same considerations, and finally you predict that this rational agent will act to further its goals in the light of its beliefs. A little practical reasoning from the chosen set of beliefs and desires will in most instances yield a decision about what the agent ought to do; that is what you predict the agent will do.*"

Dennett's concept does not depend on a specific platform, biological or otherwise. Thus, his concept was treated by the computer scientist and cognitive psychologist **Allen Newell** (1927-1992) as the knowledge level in computer systems. "*To treat a system at the knowledge level is to treat it as having some knowledge and some goals and believing it will do whatever is within its power to attain its goals, in so far as its knowledge indicates.*" The knowledge level relates to the machine selection of the best data from input on purpose, values, and information related to the problem to best fulfill its design purpose.

Evolution designed intention.

The rules of physics and chemistry apply to interactions in biochemical pathways, but those pathways' evolutionary history and rules of selection are what designed their structure to function for a purpose. The dynamics of their use, interactions, and regulation can best be explained and predicted by employing the intentional stance. Our concept of biochemical pathways as agency has been used to treat chemotaxis and also decisions to switch from glucose to lactose in E. coli. This perspective enables understanding to be gleaned through the intentional stance. This understanding brings us to the realization that evolution designed intention. Though evolution may be blind, it had purpose and was persistent in getting it right.

BDI models: artificial intelligence and living systems.

In computer programming and Artificial Intelligence, the programs for decision- making often describe agents through use of models that use concepts of cognition such as Beliefs, Desires and Intentions (BDI models mentioned earlier in this chapter). From a bio-functional perspective, the cell can be viewed as making behavioral decisions that control its ongoing biochemistry while concurrently taking into account its dynamic internal and external environments. The flow of information on states of processes is used to turn decisions into appropriate actions. A 2008 paper by Artificial Intelligence expert, C. M. Jonkers (b. 1967) and her collaborators titled *"BDI- Modeling of Intracellular Dynamics."* suggests that, *"considering a cell from the perspective of an agent sensing the environment, integrating that information with its internal state, and then choosing between possible behavioral patterns of action, may provide the basis of an alternative modeling approach."* If we view the complexity of a cell's biochemical pathways in one of the published charts ("metabolic maps"), then acquiring an understanding of behavior in terms of physical and chemical processes seems impossible. The BDI-models, however, treat the biochemical complexity as a set of stories wherein the cell is a multi-agent system of biochemical interactions with products and substrates that act as information, beliefs as knowledge, desires as motivations, and intentions as decisions. Together these lead to regulatory actions. Simulations of time series of intentions produce the dynamics realized in models of real cell dynamics. The high level of abstraction in such an analysis both simplifies programs for simulations and facilitates comprehension of the results. This is of course exactly what psychology does for neurobiology.

The biochemistry and molecular biology of a cell are very complex and dynamic. Even the simplest cell depends on 300 to 400 chemical processes that must be regulated through networked programs. The ancestral cell LUCA is proposed to have had 355 proteins. The processes of a cell have often been compared to a chemical plant that produces a broad variety of compounds for internal support, maintenance, and reproduction, while being responsive to demands for products and availability of supplies. These require webs of interactive information from inside and outside and consideration of values and priorities to continually adjust predictions about what to do next. Can we understand the management plan that makes the cell operate as a functional whole? After decades of research conceptualizing and simulating parts of the intracellular activities of a cell, in 2012 Stanford geneticist-biophysicist Johnathan Karr and collaborators published a "whole-cell" model of the bacterium Mycoplasma genitalium. This is the simplest known cell, one whose genome contains only 525 genes. E. coli has eight times and humans forty times that number of genes. Their model simulated the life cycle of the cell at the level of functional types of molecules and their interactions, starting with a newly divided cell and carrying it through to its next division. The simulation ran on a cluster of 128 computers and required about the same time as the generation time of the cell, which is nine hours. The simulation used a series of modules that mimic the various functions of the cell representing the interactions of 28 categories of molecules — including DNA, RNA, proteins and their substrates and products. Their programs organized these into modules, which communicate with one another by an exchange of data and instructions back and forth at one-second intervals. The simulated life cycle of a bacterium is another step along the way to the understanding of life. However, an indication of how far we are from understanding the roles of all genes was shown by a study done at the J Craig Venter Institute. This group synthesized a mycoplasma chromosome streamlined to 473 essential genes and found that 149 were still with unknown function. The cell is a beehive of behavioral activities just as our brain is but carried out on different platforms. Though the platforms are different they carry out behaviors with functional similarity so the behavioral models that are different in their physical implementation share common cognitive functions. Though the biochemical pathways are complex, they are simpler and more tractable in analysis so they can help us on our way to a general theory of autonomous adaptively behaving systems.

"Social lives" occur throughout the hierarchy of biology.

We usually think of social interactions as those between the same and different species of individuals. When we consider the cell as a multi-agent system, one with communication among agents that exchange information and instructions, we evoke the concept of an intracellular "social life." This concept provides the opportunity to critique the title of Richard Dawkins' wonderful book *"The Selfish Gene,"* a volume whose title is more familiar than its contents to most people. In his title, Dawkins was referring to the virtual immortality of the information in the gene rather than its manners. I have come to consider genes more in terms of altruistic team members than selfish showboats. Genes gain their longevity as a function of the success of the whole team, and cooperation thus is the key to success. As is commonly understood, genes carry information. That information becomes active when transcribed into RNAs and then translated into proteins. Less understood by most biologists are the roles of genes in the game plan that implements better behavior of the biochemical agency function. The analysis of Karr treated above involved the continual processing of regulation of gene action based on biochemical processes in phases of the cell's life cycle and intermediary metabolism.

We should not think of evolution as just competition, but also appreciate the important role of symbiosis. Cooperative multi-agency had to appear as a social aspect to intracellular regulation and between cells in multicellular organisms. We usually think of social interactions as in those between the same and different species of individuals and that also appeared early in evolution. Bacteria exhibit a wide array of social behaviors in which individuals release chemicals to organize or facilitate group activities. These include aggregation, dispersal, feeding with exozymes, symbiosis, bioluminescence, constructing biofilms, release of antibiotics, horizontal gene transfer, virulence, and aggregation for some reproductive processes. These are all activities where there must be a sufficient number of collaborators to achieve greater success and is referred to as **quorum sensing**. The earth was occupied for 3 billion years by behavior in the total absence of nerves. Biochemistry behaves at many levels of complexity. We have provided it with functions similar to beliefs, desires, and decisions. They are based on heritable memories as behavioral models that improve by error correction. They provide homeostasis through intracellular communication, overt behavior, and a social life.

A short summing-up: behavior, agency, cognition, and the role of life.

We have seen that behavior, agency, and cognition are synonyms and fundamental to the evolution of life. The goal of life is the detection of hidden causality in the world and the emergence of heritable programs that support adaptive, purposeful interaction with this world. Life as a system evolved to support itself through intervention on physical causality by directing physics to uses that are adaptive for that system. Autonomous agency is fundamental to the evolution of life and involves the creation and refinement of biochemical behavioral models to exploit environmental resources for use in homeostasis and reproduction. Life thus introduced intervention on physical causality to change what would have happened so that something else could happen. This was through evolved biochemical knowledge-based models that finessed how, why, when, and where physics could be worked to advantage with respect to the living system.

Behavior is the biological understanding of how structures can be assembled so that the flow of information and actions can exploit the laws of physics for purposes of living systems. We should regard autonomy, evolution, and agency as the product of a design process for goals, one which anticipates the future. Be prepared was the goal of evolution well before the Boy Scouts chose that motto. Evolution and cognition both use the past to predict the future. To persist and succeed, life's continual problem is — What do I do next and how can I make it work out better in the future? Living systems treat their environment as probabilistic and predict that what worked in the past will work in their future behavior. This can be modeled by inductive Bayesian inference based on accumulation of the evidence from repeated observations, where prior probabilities of fit are adjusted based on new evidence. Errors in the performance of these models must be corrected through evolutionary or developmental processes or the system will suffer potentially negative consequences.

Keeping psychology consistent with physics.

To avoid dualism, we materialists face a continuing problem of taking the psychological attributes assigned to the working human mind and translating them to be consistent with the required physics. A scientific explanation requires theories of cognitive processes such as learning, phenomenology, consciousness, free will, and downward causation to be described as having properties relating to the physics of behavior. Perhaps the best approach to understand our lexicon of psychological attributes is to analyze behavior and examine where and how these putative cognitive entities are embedded in functional roles necessary to carry out behavior. There are emergent properties attributed to agency that are new to laws of the physical world but must in some way be explained by the structure and function of the agent who develops programs to bend the laws of physics but not break them. Mind, behavior, self, purpose, value, knowledge, truth, information, error, adaptation, thought, experience, memory, perception, emotion, consciousness, quale, belief, desire, goal, intent, decision, and free will: these are only a few of the plethora of terms that we have devised as we contemplate the function of the brain. How can we treat these terms, which are widely used as mental functions but still evade reductive analysis, as recognized properties of physical systems?

Generative models: the key to making improvements in behavior.

I heard in a discussion that the new AI was based on machines that write their own software for learning and then use their performance to rewrite the software to learn more effectively. This has been the secret of life, its evolution, and the remarkable mental abilities of humans. These are based on software created by evolution and learning that can be functionally related to AI's **generative models**. These are the types of models that Karl Friston applies to living systems. *Generative learning is based on the idea that behavior in a probabilistic environment can use evidence on performance to continue to correct errors and optimize the model.* As in our introductory sentence on the new AI, generative learning involves linking new evidence with past experience to improve performance in the system it interacts with. What might this have to do with consciousness and what would consciousness be in a machine?

Something like perception and awareness can be suspected in systems that behave autonomously and seem to perform in an intentional manner.

In our discussion of Libet's half-second delay (introduced in Chapter 1), we considered the case of a baseball batter who hits a 100 mph fastball he sees approaching the plate. However, the key perceptions and decisions leading to the swing preceded their appearance in the batter's consciousness by about a half-second. This subconscious activity was accomplished through neural activity in circuits that did the necessary observations, computations, and decision-making to make the bat strike the ball. The timing of these neural activities has been measured: 100ms to see the ball, 150ms to decide the swing, and 150ms to make the swing. The retention of these events in short-term memory and the subsequent running of them through working memory allowed him to retrospectively give his consciousness credit for success. The history of the development of his skills in hitting began when his father brought home a bat and ball, and he worked his way through Tee Ball, Little League, high school, college, and the minors. As his learning progressed, his errors in timing and perceptions were corrected to get better unconscious predictions relating to the occurrence of the actual actions. As he refined his generative models of how to hit a fastball, he consciously employed coaching, practice, and introspection to improve his implicitly used model skills as they became reflexive. In the first 100ms he believed he could hit the ball, he desired to hit it for 150ms and tried to implement that decision over the next 150ms, and then kept wondering about his batting experience while trying to revive his memories. Was he unconsciously conscious of his actions in his performance at the plate since he had to observe and decide?

Scientific explanations and Sean Carroll's "poetic naturalism."

What do we expect in an explanation? How can we best treat complex systems with many participating causes that vary in their consequences and their time of action in causing some event? What do we want the explanation for? I have seen what are given as scientific explanations for ecological questions change dramatically over my career. When you read classic papers, you see that the primary reason the works are considered classics is because they set new paradigms. These often involve treatments of different levels and perspectives of approach and introduce new concepts and language. As our background understanding and perceptual perspectives change, so will our descriptions and understanding and the language and its meaning. This perspectival view underlies astrophysicist Sean Carroll's concept of **poetic naturalism,** which deals with the issue of how to objectively treat different levels of understanding in the real world. He tells us in his book, *"There is only one world but there are many ways of talking about it."* and quotes from the poet Muriel Rukeyser who wrote, *"The universe is made of stories, not atoms."* The use of the concept is treated in depth in his thoughtful book, *"The Big Picture: On the Origins of Life, Meaning, and the Universe Itself."*

The tenets of poetic naturalism can be summarized in three relatively straightforward points, given below. In considering these points, it is important to keep in mind that analyses using poetic naturalism to explain a particular level of reality must not violate or reject evidence at other levels of understanding:

1. *"There are many ways of talking about the world.*
2. *All good ways of talking must be consistent with one another and with the world.*
3. *Our purposes in the moment determine the best way of talking."*

David Marr: computational, algorithmic, and implementational levels of understanding behavior.

What is to be explained in behavior? Earlier we introduced Dennett's physical stance, design stance, and intentional stance as ways to explain behavior. The British neuroscientist **David Marr** (1945-1980) employed perspectives from psychology, neurobiology, and AI in models of visual processing. His diverse perspectives on visual behavior suggested three alternate but complementary levels of understanding: the *computational, algorithmic,* and *implementational* levels. These approaches have become accepted broadly, and Marr's approach remains an intuitive framework that can be used generally to treat models of cognitive and perceptual systems. These levels in living systems are recursively interactive in their formation of output. Action introduces a change of input and the next computational problem starts all over again, as long as the system keeps running.

"COMPUTATIONAL: *At this level we describe and specify the problems life faces.*" (This is the level of required response to homeostatic and environmental issues.)

"*ALGORITHMIC: There must be a software that structures a knowledge link between the computational and implementational levels, describing how the identified computational problems can be solved.*" (This is the level of adaptive behavioral models for cognitive causality.)

"*IMPLEMENTATIONAL: Physics requires the hardware as the physical substrate of mechanism and organization. The algorithmic level designs the hardware by which a computation is achieved.*" (This is the level of building physically causal models.)

Each level should be seen as a realization of the level before it. Thus, the algorithmic level is a response to the computational level, describing how the computational problems can be solved. Similarly, the implementational level is a realization of the algorithmic level, specifying the mechanism that carries out our algorithms. Marr introduced these levels as a reaction to arguments between disparate research paradigms. Instead of emphasis on implementational or computational problems, he examines algorithms that bridge these concerns.

The free energy principle, Bayesian inference, and generative models have a fit to Marr's framework, and these points are shown in bold font below. This listing can help to generalize to other research efforts. Included are Dennett's three stances of explanation and the BDI approach to psychology, both of which fit into Marr's framework. All of the terms used in discussions sort out into these levels, as we document below.

Computational: (function, purpose, value, Problem, teleology, final cause) **Friston's purpose, adaptive value and goals of generative models and includes self and observation.** [Desire in the BDI of psychology and Dennett's Intentional Stance.]

Algorithmic: (model, program, Software, design, sub-functions, knowledge, information, properties, meaning, representation) **His program of the generative model, Bayesian inference, information, knowledge,**

representation, and error correction in model design. [Belief in psychology and Dennett's Design Stance.]

Implementational: (physics, platforms, Hardware, molecules, matter, energy, data, symbols used as information flow in perception and action, behavior) **This is the physics of the structure and the behavior of the generative model.** [Intention in psychology, and Physical Stance of Dennett.]

Marr's problems and solutions appeared and operate in the cognitive realm with the emergent properties of living behaving systems that are cyclical, i.e., recursive in their interactions. Thus, these living systems behave recursively to implement the physical structure that then behaves and is observed reaching goals. Marrs' purpose, design and matter combine for success of the entities that interact physics with information used for cognition. This includes living systems, but his approach is general for autonomous adaptive systems that behave. Consciousness contains psychological functions that relate to the role of an observer. These functions involve perception, meaning, and value of an intention, which are combined in Marr's first two levels. In our model's operation, consciousness can be viewed as the free energy, which is monitoring interaction and observes how well behavior is going and what to do to improve it. How does the free energy relate to representation of how the behavior feels? The free energy principle is a stand-in for the surprise function in belief versus evidence. In summary we see these four approaches to behavior are all based on a physical structure that is designed by an algorithm for an adaptive purpose and we treat this as the basis of life.

Cognition is represented in the structure and actions of the physical entities that we study as biological systems. This implementational level constitutes what represents a reductive scientific explanation. When a classical physical system is analyzed, that is as good an explanation as we can get. However, in biological systems purpose and design are causes of the physical implementational level. They are necessary for understanding the how, why, when, and where of living systems. Chemistry in vivo may look like chemistry in vitro, but without the cell it would not happen. The implementational level in biological systems cannot be totally explained by physics since life has design for a purpose with meaning and value, which are not in the domain of physics

Information emerged with life by its use as tokens or signs of physical entities in an agent's environment that both structured the knowledge in its behavioral machinery and guided its interactions. Though we can use this information in our stories of behavior, the evolutionary causality for its use as signs is lost in a morass of eons of historical contingency. *The meaning of the information is bio-functional and relates to the heritability of developed knowledge programs.* There is no red in the real world; it only exists as neural activity in your brain. Red as information represents electromagnetism as a structure of activity in a neuronal assembly. We assign color as a property of an apple as one of the reasons why, how, when, and where the physics of eating one happens. Building the knowledge program of apple eating behavior assigned this information sign of color along with other properties relating to shape and taste as stimuli to act behaviorally. Essentially, knowledge and meaning converted data on physical characteristics of the apple into information incorporated into the knowledge model in a clever circular interaction that is not predictable by the rules of physics. What does an apple look like to a deer or worm?

For living systems to persist, reproduce, and evolve they were compelled to discover and exploit hidden sources of causality in a complex, dynamic, and opaque world. Life is based on exploiting resources by the gradual emergence of observation, induction, hypothesis formation interacting with models, and their intervention on the physical world optimized by feedback used for error correction. This is **approximate Bayesian inference** and the pathway needed to build behavioral systems in both biochemical models and neural networks. There has recently emerged the recognition that perception, learning, and even evolution could be treated with Bayesian statistical models. The Bayesian brain hypothesis uses Bayesian probability theory to formulate perception as the brain optimizes its model of the world using sensory inputs and behavioral success. Thus, the brain is an inference machine that actively predicts and explains its sensations with probabilistic models that can generate intentions for action. These models are optimized in their action by using errors in perception to update belief about their cause. Karl Friston provides formal mathematical treatments of our predictive brains performing active inference, based on the use of generative models. Consider this analogy: All the notes for all possible music were pre-existent in the world, and the patterns and time of their incorporation into compositions has produced the emergence of all of our music by selective processes in different environments. Life has done this with molecules crafting them to compose symphonies of biochemistry.

We introduced a profound statement in the discussion in the paragraph above: *The brain is an inference machine that actively predicts and explains its sensations with probabilistic models that can generate intentions for action.* Most of what life does has been refined and incorporated in old and well-used models that have been designed for a specific adaptive response. Thus, their desires and beliefs have predesigned intentional paths. Our capacities span from reflex withdrawal from something hot, to knowing a pattern of muscle contractions needed to pick up a cup, to playing a selection on a piano. Compare these behaviors to a baby trying to pick up a cup or a student just learning the piano. These behaviors have been acquired through pathways of evolution, development, and learning where the wiring became more or less hardwired through repetitive use and selection. These account for virtually all of our behavioral control that goes on in the unconscious background to our consciousness. Elements entering our consciousness are the new observations, ones that attention will refer to pathways in order to produce perceptions, which then can call for decisions concerning actions. This has generally been considered as input proceeding from the bottom up for perception and from top down for decisions. This is now contested by the **predictive processing concept with the brain as an inference machine.**

We have stressed that a complex and probabilistic world calls on Bayesian inference to make predictions of how good our guesses are concerning input and what we should do about it. We use bits of memories from our past to recognize and predict how to respond. In variable and free form environments, we must improvise responses that are adaptive. The raw perceptual inputs can be compared to the memories of similar experiences. Input data that are like previous input proceed through the most familiar paths as established by facilitated synaptic connections. Input gets directed on pathways according to our refrain, neurons that fire together, wire together. These similar experiences in the past establish priors used to assemble components of a beginning generative model. This produces the first rough cut at a model based on the similarity to existing knowledge that feels like the best guess. The model is recursively compared when fed back and judged for its accuracy. It is the error in model performance that becomes the information sent up to correct and achieve the best fit for the purpose. This is of course the theory of associative learning, which comprises practice and error reduction to better predict how to achieve a goal. This use of prior knowledge makes the richness and degree of thematic cross-referencing that has preceded the new experiences of great importance. Of course, the truth-value of the knowledge used for priors plays a major role in the validity of the generated model. To summarize: life can be seen as a continuous learning process, one that has passed on 3.6 billion years of the memories for the essential processes that we employ to keep living and cope with ongoing challenges and opportunities.

The American psychologist and philosopher **Alison Gopnik** (b. 1955) introduced me to the concept of babies behaving as scientists. Thus, during their development, babies acquire their view of the world through processes similar to those used in experimental science: observation, induction of hypotheses, and refining of these hypotheses through more observations, experiments, and error correction that employs Bayesian inference. She also reflected on scientists continuing in the ways of children. Unfortunately, many individuals give up the search for evidence to correct their beliefs once they grow out of childhood. Since life developed knowledge models to intervene on the ongoing physics, we can conceptualize life as analogous to science where species work to make models of causality in order to exploit their environment and survive in their world. Behavior analogous to science has worked to life's advantage to solve problems for billions of years and produced the human species that formalized science and exploited its products. The value of science in solving emerging problems for society must be recognized. Behavior, purpose, goals, values, and meaning all are functional evolutionary emergents (emergent properties) of living systems, which have grown evermore complex. If living systems with agency, behavior, and the properties of cognition arose as an intervention on physical systems, it should not be surprising that the laws of physics cannot provide a complete explanation of life processes, especially those of the mental realm.

Further Reading and References.

Campbell, J. O. (2017). Universal Darwinism as a process of Bayesian inference. Frontiers in Systems Neuroscience 10: 49. https://doi.org/10.3389/fnsys.2016.00049

Carroll, S. (2017). *The Big Picture: On the Origins of Life, Meaning, and the Universe Itself.* New York: Dutton

Dennett, D. C. (1989) *The Intentional Stance.* Cambridge MA: MIT Press.

Norton, B. G. (2005). *Sustainability: A Philosophy of Adaptive Ecosystem Management.* Chicago: University of Chicago Press.

Gopnik, A. (1998). *The Philosophical Baby: What Children's Minds Tell Us About Truth, Love, and the Meaning of Life.* New York: Farrar, Straus, and Giroux.

CHAPTER FIVE

Mind-Body Duality & Causality

The Mind-body problem: the basic challenge.

When we speak of the relationship between mind and body, we are contrasting properties of our mental world and those of the physical world we inhabit. From the perspective of science, we typically feel that explanations should reduce to the physical properties and their laws, which have created our shared and agreed upon understanding of how our physical world works. Since our bodies are made of physical stuff, the mental properties therefore must reduce to this physical substrate or become mysterious. The result of this position is that the mind must have emerged from the properties of the body at some time in the history of evolution. Introspection of our own minds and thinking about how this relates to other minds was why we assigned mental properties and placed their emergence in complex brains.

The central problem related to minds is how neural biochemistry, ion fluxes, and squirts of chemicals could produce the *feel* of consciousness and thoughts. Do feelings occur with the neural activity of behavior, or do feelings only happen in a specific site. Does perception have a feel? What did driving feel like when you drove some distance while unaware of your actions? There is a half-second or more delay of neural activity before it becomes conscious and our memories are reruns of experience. The reportable feel of an experience may only be a replay of information about the experience in **working memory** as we review it. Remember details of our batter and the 100 mph fastball in the previous chapter. Or is the feel part of the experience? Could the batter have 'unconsciously' felt very similar on the initial experience but had to wait a half second or much longer for the review. On review the batter takes credit for initiation and control of the conscious creation of the experience. The subjective first person experience of consciousness has dominated research and concepts of the functional role it plays in behavior. Human introspection, culture, philosophy, and science have focused on explaining our human concepts of consciousness only as an emergent phenomenon of higher order neural action. This focus has produced impediments to understanding the function of consciousness and other mental processes as evolved adaptations in living systems.

Co-evolution of mind and body.

With a lot of help, I have reached the opposite perspective, namely that bodies of living organisms have emerged from interactions with their cognitive abilities. This line of reasoning proposes that knowledge, memory, information, purpose, values, perception, and adaptive error correction were central to the structuring and activating of all bodies. Living bodies are structured biochemical actions with a purpose. In other words, they are behavior with properties of agency. Agency entails cognition and is based on similar functional properties to those of behaving minds. This view leads to our concept that cognitive properties emerged during evolution and turns the function of consciousness into a guide adapting life to variable environments.

Consciousness is treated in many forms and functions and how and why it feels is left to introspection. Consciousness aligns with concepts such as the following: what it feels like, quales, properties, experience(s), thoughts, perception, representation, observation, and measurements. I find the last three terms functional in behavioral models that must represent the modeled system. As cybernetic pioneers, Roger C. Conant and W. Ross Ashby (1903-1972) tell us, *"Every good regulator of a system must be a model of that system."* This idea was fundamental to life acquiring resources where the physical events that could provide resources could be modeled in a perceptual interactive framework. Observations are made; data are transduced to information. The information is then supplied to construct the model, which, in turn, is recursively shaped and corrected by behavioral interactions into adaptive function(s). Consciousness would fit the roles of observation, control, and error correction as the director of our behavioral actions and also be employed in writing new scripts.

Supervenience: is physical structure supervenient on cognition?

Physical properties and laws, which we all can share and accept, are involved in living systems. A long standing position in philosophy has placed the mental properties as **supervenient** on their physical substrate. This poses the questions that must be answered: Does physics tell us what to do? **or** Do we tell physics how to behave? My book is about life intervening on physics and instructing it in why, where, and when to act as it behaves. No laws of physics are broken by living systems, rather they are implemented for a purpose. Information, knowledge, and consciousness (representation or perception) are emergent with agency. We call this functional **emergentism**, where the physical structures and energies are **supervenient** on cognitive behavioral programs implemented in the generative model. Recall that the concept of supervenience states that X (here, the physical structures) is said to supervene on Y (here, cognitive behavioral programs) if and only if some difference in Y is necessary for any difference in X to be possible. The dualists had looked for just the opposite supervenient relationships: supervenience of the mental (Y) on the physical substrate (X). I realize my proposal sounds circular since knowledge and information require a physical substrate. However, the origin of life supplied the structure as a replicating molecule that had to compete for resources. Replication of variations that happened to be of value, contingency, and different pathways of Bayesian inferential optimization have led to life's current diversity. In a living system, agency, its physical substrate, and energies are networked in cognitive control architecture. Evolution and learning recursively create functional behaviors in a dynamic setting. They must be adaptive, regulated, and error correcting systems that are interactive with their setting.

Emergence: life brings causation, agency, and purpose to the physical world.

Stuart Kauffman regards the emergence of life as the formation of heritable biochemical agency capable of implementing work cycles for its maintenance and replication. He thoughtfully states in, *Beyond Reductionism: Reinventing the Sacred*;

> *"We have lived under the hegemony of the reductionistic scientific worldview since Galileo, Newton, and Laplace. In this view, the universe is meaningless, as Stephen Weinberg famously said, and organisms and a court of law are "nothing but" particles in motion. This scientific view is inadequate. Physicists are beginning to abandon reductionism in favor of emergence. Emergence, both epistemological and ontological, embraces the emergence of life and of agency. With agency comes meaning, value, and doing, beyond mere happenings. More organisms are conscious. None of this violates any laws of physics, but it cannot be reduced to physics. Emergence is real, and the tiger chasing the gazelle are real parts of the real universe."*

The following section may well be the most counter-intuitive and controversial of my book. The concept of an emergent universe, one that has produced the creativity of evolution with purpose and adaptation, introduces concepts of cognition and inserts them into physicality. We have already given thanks to Sean Carroll for his introduction of poetic naturalism that facilitates discourse at different levels of emergence. The only restriction for meaningful treatments in using different languages to discuss different levels of emergence is maintaining consistency between the levels. Properties used at an emerged level must be an extension rather than a violation of the rules at the other levels.

Before I was introduced to Sean Carroll's perspective, I thought of causation at the level of fundamental interactions in the physical world and the role of life was to discover and use hidden causation. Since causation is so important to both life and our explanations, I have read on it extensively, especially in my search for understanding of causality of mental properties. How could the concept of mental causation be emergent from the underlying physics of a living agent? I encountered a paper by Sean Carroll titled, *"Why*

(Almost All) Cosmologists are Atheists," where he told me that I could view causation as an emergent property with life. In the quote below, I added the bold as an emphasis that evolution and the emergence of life is that **additional layer** and the introduction of causation. Purpose and causation were emergent with life and its quest for resources to maintain itself.

> *"From the perspective of modern science, events don't have purposes or causes; they simply conform to the laws of nature. In particular,* **there is no need to invoke any mechanism to "sustain" a physical system or to keep it going; it would require an additional layer of complexity for a system to cease following its patterns** *than for it to simply continue to do so. Believing otherwise is a relic of a certain metaphysical way of thinking; these notions are useful in an informal way for human beings but are not a part of the rigorous scientific description of the world. Of course, scientists do talk about "causality", but this is a description of the relationship between patterns and boundary conditions; it is a derived concept, not a fundamental one. If we know the state of a system at one time, and the laws governing its dynamics, we can calculate the state of the system at some later time. You might be tempted to say that the particular state at the first time "caused" the state to be what it was at the second time; but it would be just as correct to say that the second state caused the first. According to the materialist worldview, then, structures and patterns are all there are — we don't need any ancillary notions.*"

Downward causation, which has worried physically-oriented thinking in philosophical discussions of the reality of mental properties, would then disappear as a problem. There was no causation before life appeared and intervened on process in physics. Sean Carroll informed us that,

> *"of course scientists do talk about "causality", but this is a description of the relationship between patterns and boundary conditions; it is a derived concept, not a fundamental one."*

Boundary conditions establish limits or constraints on the system being analyzed and this applied to life. Life supplied the boundary conditions and causality emerged through intervention on the physical system, which was necessary to sustain life.

Studying causal relationships helps us understand physics. Our ability to communicate about physics has resulted from establishing boundary conditions that allow us to study and make measurements of study systems. Cause effect analysis has become enshrined for physical explanations and also the criterion for the sciences. During the history of physics, boundary conditions were established at successively lower levels to promote deeper understanding. The documented success and usefulness of physics has placed its method of reductive explanations as the accepted paradigm in science and philosophy. In our desire to remain physical and not invoke mysterious qualities and actions, science based explanations on the machinery of behavior rather than its origins, purpose, and control. What happens to an explanation when causation becomes multiple, in part historical, dynamic with conditions, and produced by complex machinery? The analysis of complex neural behavior may well fall into this category of being refractory to a reductive causal examination. Thus, general theories of behaving systems become important. Behavioral functions and properties from comparative approaches that examine different levels of complexity can provide tools for a general science of cognition. Understanding cognitive functions as programed behavior of biochemistry provides perspective on the mystery of process and function of cognition in life.

The mystery is made less opaque by anchoring life's origin with the view that causality emerged with life. This view provides the opportunities of dynamic and adaptive causality. **Counterfactuals** are the where different actions or goals can be selected and this occurred with life's intervention on the physical world for energy and resources. One approach to causation sees counterfactual dependence as the key to the explanation of causal facts and there are centuries of philosophical discussions of causality.

Here, we have introduced the concept that living systems exist through having established boundary conditions where they could depart from the inevitable interactions of matter and energy established by physics' rules. This departure from strict domination by physical rule is directly tied to the behavioral analyses of neuroscientist Karl Friston (first mentioned in Chapter 1). Friston establishes generative models for adaptive behavior following Bayesian principles by correcting errors, which he quantifies as free energy.

The free energy principle for agency depends on the presence of a **Markov blanket** as the boundary conditions between the agent and the world it interacts with. A Markov Blanket partitions a system into internal and external states, where the sensing and action of active inference pass through the blanket. In biological agents this applies to the model during active inference. In an individual these blankets are multiple, hierarchical, and changeable. A Markov blanket links models in interactive behavior with resources only by permitting flow of information and action. Thus, the emergence of information with meaning for adaptive behavior was necessary for life.

Biosemiotics and symbolic systems: information, communication, and life.

Biosemiotics is defined as, "the study of signs, of communication and of information in living organisms." This study of signs permits the 'intentionality' of biology to interact with other systems. In a 1969 paper titled, *"How does a molecule become a message?"*, in *Developmental Biology Supplement* the American biologist Howard Pattee (b. 1926) reflects that in evolutionary systems, molecules are often messages transferring behavioral information involved in interactions with another system. These molecules thus are not acting purely physically but are part of an emergent system of agency. The function of conveying a message introduces new concepts of information, knowledge, meaning, value, goals, and purpose, and the assumption of the detection and translation of signs into symbols for communication. Concepts can be related to the experiencing of observation (representation or perception) of both external and internal systems. The early insight of biosemiotics was that living systems, from biochemical pathways to organisms, are controlled by information, goals, and values.

Organisms interact with the physical world as sensorimotor systems. Energy from the physical world comes in contact with sensory surfaces, and interactions are based on how this is transduced into symbols with meaning for a behavioral model. This is the basis of Friston's generative models formed through active inference by minimizing free energy. Each cognitive model is interactive with a set of these symbols to which it assigns cognitive value. Symbol systems have become the tool of choice for studying language and cognitive systems and provide the additional advantage of being approachable through computer simulation. A symbol system is a set of arbitrary "physical tokens" that can be manipulated by rules encoded in the behavioral model.

These systems can be recursively modified to interact more effectively with the environmental model.

Knowledge and meaning relate to a purpose or problem. They are formed by data becoming information that is provided meaning related to the knowledge level concerning solution to the problem. Knowledge and meaning emerge from the success (or failure) of the model's predictions of success in solving the behavioral problem. The free energy minimization is the recursive selection of the programs hardware to enhance model performance. Information, knowledge, and meaning emerge recursively linked in evolution and learning. Effective competition for behavioral goals selects physical structures that symbolize information, These structures intrinsically represent information, knowledge, and meaning achieved through the model's history to predict performance. Life has employed symbols as functions for cognition which represent mental states associated with their functioning. This perspective proposes how meaning, consciousness, and feelings arise as a complement of the physical platform of autonomous evolved adaptive behavior. Physicists tell us that an electron is both a particle and a wave depending on how it is observed. We can observe behavior as its physical implementation or observe the associated information for the problem and algorithm. Biosemiotics can justify cognitive properties in evolved biochemical pathways but of course the question of cognitive properties in machines remains as a question to be resolved.

Perhaps going back to the question of whether a thermostat is conscious will provide insight here. In a thermostat, the control function is designed to provide adaptive results for the designer. The algorithm is supplied by the designer and the sensed set points are selected by the way it feels to the user. Our body's thermostat is based on an algorithm designed by evolution to provide adaptive behavior for our homeostasis; thus, the sensory function is intrinsic to how it feels to the behaving system. Finding any problems with desired goals and solving these problems is the process that characterizes generative models. This process involves the individual as the sensing self that uses how it feels to get the design right. From bacteria to humans, thermostatic mechanisms have appeared for selecting optimal temperature conditions. Active inference recursively works to build causal models of the world, which life interacts with for resource and energy. These causal models provide our knowledge and meaning of the world we interact with for opportunities and challenges. Consciousness as perception is the observation of the action of the models' predictability for gaining behavior's success. The value is how it feels related to the relevance of the behavior's purpose.

These relationships were what we explored in the behavioral biochemistry and chemotaxis in *E. coli* at the level of microbial cognition. Remember, as Harvard physics and biology professor Howard C. Berg defines it: *"E. coli, a self-replicating object only a thousandth of a millimeter in size, can swim 35 diameters a second, taste simple chemicals in its environment, and decide whether life is getting better or worse."* This quote on *E. coli*, when combined with quorum sensing and social behavior of bacteria, provides a description of a behavioral pattern of life on earth for over 3.5 billion years. Whether simple biochemistry, like that discussed for *E. coli*, or neurons and brains associated with metazoans are involved with decisions and choice of path, life must model the world it interacts with for its resources. As Stanford developmental biologist H. Pattee reminds us in his 1969 article, *"How does a molecule become a message"*, this was first accomplished by using molecules as messages in behavioral pathways. Chemistry *in vivo* may look like chemistry *in vitro*, but when outside of the cell it is just the physics, which is not a complete explanation of what the chemistry accomplishes when it is inside the organism. As materialists we believe scientific beliefs must be consistent with physical laws. Studies of living systems discover no violations of physical laws but instead reveal ways in which physics embraces new rules.

The origin of causation by life.

Before life stuff just happened following the rules of physics. There was no causation and no unique reductive explanatory causal chains before evolution inserted intervention and selection into the unfolding of life. Adaptive causation emerges from cognitive functions of behavior exploiting opportunities. These functions are incorporated into generative models as the observing cognitive system is shaped to adaptively interact with other systems. In a causal analysis, contrary to reductive physics, causal arrows point upward, downward, and sideways. We have created our scientific explanations in the dominant perspective of physical reductionism but live our lives under the regulation of evolutionary cognitive rules, which will not break physical rules but will exploit them to the individual's advantage.

We have spent much of the last century trying to demonstrate that life itself is nothing but physics and chemistry and how it lucked out in finding

DNA. The results of research provided detailed descriptions of the structures and energy changes of life, as we unraveled biochemical pathways, molecular biology, and their roles in life processes. One could presume this chemistry along with the lucky accidents encoded in genetic information that favored reproduction would take us up to complex animals. Their origins would provide a nervous system and the miracle (emergence) of consciousness along with subjective feelings and thoughts. The brain seemed to work with rules that defy explanations based on the primacy of physical causation. I would claim our causal descriptions of physics were dependent on life evolving as a cognitive system requiring behavior. It was life that derived the concept of causation. Life provided useful properties to those aggregates of fermions and bosons bouncing around in the enigmatic world emergent from the quantum realm. As life sought to understand and predict physics for its behavior, causation with counterfactuals arose on earth. Observation, representation, and perception of the environment supply information for the behavior that provides for our needs and goals.

Consciousness has always been in search of a definition to tell us what it is and what it does. What is the relation of consciousness to causation? Our cognitive model's recognition of the system with which it interacts is necessary to elicit behavior. Is causation the act of creating information for representation? Is causation in action a definition of consciousness? To explore these issues, let's examine the process entailed in a belief-desire-intention (B-D-I) behavioral action model. On page 11 of Chapter 4 Marr's 3 levels of behavior were aligned with properties of Friston's generative model, psychology's BDI model and Dennett's three stances to show their similarities. An organism confronts challenges, needs, and opportunities that require behavioral response. A flow of sensory information activates models with desires to solve the problem and beliefs that their actions will accomplish this. Action is initiated that must be recursively controlled and corrected for success. These are available by introspection as the commonsense perspective of consciousness and causation. They fit the functionality of behavior implemented by a generative model in action. The generative model represents the cognitive behavioral program assembled to interact with the targeted system. The two systems are separated by the Markov blanket, which functions to allow the inward flow of information and the outward flow of action. The outward flow of action can result in an event in the environment, which can be interpreted as information by the same or another agent. The process causally links behavior and its cognitive control in a recursive manner.

How, why, when, and where the behavioral system acts depends on the flow of information. For the information to flow, the system must harvest energy from the environment, which, along with harvested matter, is required to maintain and reproduce itself. This process is ongoing in the behavior of life, and, if the behavior stops, the living system reverts to just physics. Consciousness as function creates information with meaning whether we can report it or not. Thus it is also employed by us in implicit learning that goes on below our awareness. Observation, representation, perception, and control all require the function and who knows how it feels in any other cognitive system?

Consciousness becomes the feeling of the knowledge, information, purpose, and value when information flows in action, as observed by the behavioral model. We have stressed that properties emerging from the structure of the model have their ultimate function as cognition in behavior. Correcting errors in behavior requires adjusting the physics of the action or the model to optimize model performance. Errors represent the prediction error of the model. This is reflected in utility, reward, value, or the feeling of how life is doing in its behavior, which is important because errors in behavior can threaten life. Human behavior's complexity in the underlying machinery of nerves, synapses, neural transmitters, use of memories, contingencies, and the associated feelings create problems for scientific explanations. In physics, scientists have faced similar problems in understanding how the actions of a quantum world appear as classical physics and coping with the difficulties of reductive causal explanations. Since we share a similar problem in the brain, perhaps we can share solutions.

Effective Field Theory: Common strategies in physics and biology.

Effective field theory was necessary to treat emergent properties of classical physics from the micro high-energy world of quantum mechanics. Field theory justified formulating laws based on large-scale emergents from the statistics of micro-aggregate actions. In physics, an effective theory is used to treat some larger scale system as an emergent approximation to a deeper theory. An effective theory models the features of the system that are important to you at the level of interest. These are often larger scale features that are statistically produced from reductionist microscopic properties' interactions. Effective field theory was developed to deal with lower energy emergents of the quantum world but can be broadly applied. Things such as protons and neutrons are effective theories of quarks and gluons. Atoms and the Periodic Table of elements arise from protons, neutrons, and electrons. The gas laws are a statistical macroscopic treatment of the collisions of molecules in a gas. General Relativity is generally accepted to be the low energy effective field theory of a full theory of quantum gravity that is still searched for. To conceptualize General Relativity, the American physicist John Wheeler (1911-2008) told us that, *"Matter tells spacetime how to curve, and spacetime tells matter how to move."* It is a circular process without a reductive causal chain. We hope to supply more understanding as physicists continue their search for quantum gravity.

In this chapter, I earlier posed the questions; Does the physics tell us what to do? **or** Do we tell physics how to behave? Perhaps my consideration of these two questions is how I have reached my understanding of consciousness where, like gravity, I lack a metaphysical metaphor to help conceptualize 'what is there'. To help with the rationalization and reality of the emergent cognitive properties, an effective theory can be applied. Let me paraphrase Wheeler's statement (above) on the enigma of gravity: *Life tells physics how to behave, and physics tells life how to feel.* Their interaction provides me with a functional, naturalistic, and non-mysterious understanding of the emergence of new properties for the control of life. These properties arise from the elements of physics when they are placed in their new functional relationships by living processes. For me then, the emergent properties that go into the control of the physical realm are the feel of the flow of information to effect knowledge, purpose, and value. Since none of this existed before the appearance of life, I would propose consciousness introduced causation in the origin and optimization of behavior. Intervention and counterfactuals (from page 5) require observation, which supplies something like representations or percepts for adaptive behavior. In life this lacks the formal predictability of the physical world since in biology this all occurs in individual systems each separated by its own boundary conditions or Markov blanket as discussed on page 5. This introduces individual histories with contingency and context driving causalities and different pathways being taken. Counterfactuals have emerged with life and different and unpredictable things can happen.

The section above covers the consciousness associated with the operation and optimization of the generative models of behavior in life. Most of those who treat the subject of consciousness are only concerned with its appearance in autobiographical memory, where it can be talked about. However, all the process necessary to create and relate these stories along with their feel is dependent on their production in the unconscious. In a conversation, I am only aware of what I will say when I hear myself say it. People think more about what consciousness feels like to them instead of asking what it could do as an evolved property of life. There will always be the mystery with respect to "what it feels like", as it occurs in others and as a general process in all behavior. What makes consciousness in life difficult to explain is the span from molecular biology to human thoughts and emotions.

Sean Carroll in *The Big Picture* points out two features that can provide both generality and breadth of explanatory power to an effective theory. *"For any one effective theory, there could be different microscopic theories that give rise to it."* Thus, there is an independence from platform, and this means we don't need to know all the details of the physics to make statements about macroscopic behavior. This independence allows cognitive psychology of humans and behavior throughout life's diversity. A cognitive effective theory allows control theory, free energy and generative models, agency, robotics, and AI, to all be treated as their physics designed for adaptive behavior. This does not depend on a special region of the brain or even a neuronal platform. The second point was, *"That for any effective theory, the kinds of dynamics it can have are generally limited."* This is borne out by our comparison of the variety of approaches to behavioral studies that all align to adaptive behavioral agency that recursively interacts under the shared rules devised early in the evolution of life.

Consciousness in organisms and in machines.

Devotees of David Chalmers' 'hard problem,' wishing for a familiar metaphor to get to the metaphysics of their stories of personal experiences, will remain disgruntled at this level of description. I would say that we are filled with feelings we simply can't report on. Remember the reportable conscious works at 40 bits per second. Humans encounter a mental block on the concept of feelings in biochemical pathways, nerve nets, and perhaps even a greater rejection for machines. When I began my latest attempt to understand cognition and consciousness, I found the approaches of those in computer processes, artificial intelligence, robotics, and artificial life provided new perspectives. For those unhappy with feelings in biochemistry will be very upset to consider how silicon chips feel. The two features of effective theories introduced above are powerful in bridging the evolutionary and AI linkages in a general theory of behavioral control and their functional processes. Poetic naturalism provides common language in a general theory. We can speak of the properties of knowledge, information, purpose, observation, value, representations, consciousness, and decisions in adaptive behavior that we don't find in physics. There was a huge activation energy hump for me to crawl over since they use a different language in AI and robotics. My struggle was to create natural history stories from the AI and robotics literature.

Consider driverless cars, for example. Fiction writers may be better prepared than scientists and philosophers to recognize consciousness in the Google's Waymo self-driving car program that has driven autonomously over ten million miles. If you hailed such an autonomous vehicle as a taxi and it was equipped with an enhanced Siri, topically oriented chatterbots, the knowledge of IBM's Watson, and the motivation to converse, then the belief of consciousness for the vehicle gets more believable. Consider owning such a vehicle that learned your interests and adjusted its programming on the basis of years of interactions with you. It might become one of your best friends, and perhaps even share stories about your shared experiences as it gossips with other intelligent vehicles.

This comparative framework provides a general and definable description of operations of mind in an agent so it can be compared over a broad range of platforms and complexity levels. This synthesis brings together the evolutionary-to-robotic spectrum of behavior and mind properties, allowing all of them to be examined for continuity of process and their related subjective descriptors, to see if and when the emergence of mysterious non-measurable properties is required.

We have spent this chapter creating a story of how life emerged from our physical world. We have treated behavior as cognition from perspectives of philosophy, biochemistry, neurobiology, psychology, ethology, robotics, and AI. Throughout much of its history, physics consisted of separate stories concerning different approaches to its study. This quote from page 256 of British mathematician and historian Jacob Bronowski's (1908-1974) book, *The Ascent of Man*, emphasizes the importance of how everything must work together for consistency and coherence to make a story complete.

> "So in a lifetime Einstein joined light to time, and time to space; energy to matter, matter to space, and space to gravitation. At

the end of his life, he was still working to seek a unity between gravitation and the forces of electricity and magnetism."

Just as Einstein revolutionized thought in the field of physics, those in cognition must strive to join emotion to attention, and attention to consciousness. This quest must begin with assigning values to information, using information to infer hypotheses, representing hypotheses by probabilistic networks, selecting salience for adaptive decisions, reviewing decisions in consciousness, and using consciousness to feed back values and probabilities to optimize future output. Bayesian processes are the solution to optimization of cognitive process and the behavior it produces. As the formal theories that treat cognition blend with the approaches in AI and robotics, we begin to translate the subjective into the objective and anticipate the future.

Making the best decisions—what evolution is all about and what will determine our future.

Evolution and development inform us that what we learn becomes increasingly dependent on what we know. The more prior models of knowledge, and the better and more diverse they are, the more an individual can use them to discover and process new and surprising information to keep adaptively abreast of environmental change. In a complex and rapidly changing world, a shortfall in questioning of beliefs becomes the overwhelming problem faced both at the level of the individual and societies. This problem becomes of crucial importance when those less informed and unwilling to learn make decisions that affect others. As this persists it can create problems that threaten our future. We have seen that evolution got us here by using history to predict the future; for any species to survive it must innovate responses to a complex and dynamic environment. Those species that didn't do this were dead ends, while the perceptive and responsive ones were ancestral to successful lineages. During 3.8 billion years of evolution, success in predicting the future provided us with brains that natural selection shaped as observing, inference generating, prediction-making machines to get it right. Effective use of these capacities becomes a critical issue when our predictions are used to structure a suitable environment and we get it wrong.

Much of what we intuitively believe about our decision making process differs from how our neurons do it. This is supported by measurements of bandwidth difference between our consciousness, which we think we use to run our lives, and the unconscious, to which we now only assign more reflexive influences. However, consciousness works at a bandwidth of around 40 bits per second while the rest of the brain processes at 100s of millions of bits per second. Consciousness is also delayed by around a half second or more from the neural activity that determines perception and our decisions. The paucity of content and its delay in becoming conscious, which we are not aware of by introspection, nevertheless heavily structures our behavior. In our introductory chapter we reflect on what should be the role of the conscious mind in making decisions. Since the way to improve performance in behavior is through practice and correcting errors, this should be a primary function of consciousness. In order to correct errors, they must be discerned as a difference between your belief and reality. This corrective process presents the challenge of using all the evidence to create and update the probability of your belief.

Further Reading and References.

Carroll, S. B. (2016) *The Big Picture: On the Origins of Life, Meaning, and the Universe Itself.* New York: Dutton.

Carroll, S. B. (2005) "Why (Almost All) Cosmologists are Atheists," Prepared for God and Physical Cosmology: Russian-Anglo American Conference on Cosmology and Theology, Notre Dame, January/February 2003. Published in *Faith and Philosophy* 22, 622.

Conant, R. C., and W. R. Ashby. (1970) Every Good Regulator of a System Must Be a Model of That System. Roger C. *International Journal of Systems Science* 19: 89-97.

Pattee, H. H. (1969). How does a molecule become a message? *Developmental Biology Supplement* 3: 1-16.

CHAPTER SIX

Deep Learning and The Default Mode Network

"In the end, we are self-perceiving, self-inventing, locked-in mirages that are little miracles of self-reference." I am a Strange Loop — Douglas Hofstadter

Deep neural networks and information bottlenecks.

At one time I was a regional authority of identification of enigmatic limpet species. This skill took decades to develop and entailed the consolidation of evidence and its selection from viewing over 1/2 million limpets. It seemed that the criteria I used for the tough ones were implicit and not available explicitly to communicate to others. I had no idea how I sorted evidence and was unable to adequately communicate all my decision criteria to others. I have recently been informed by my reading on deep learning in AI that complex learning may have some similarities in learning from huge amounts of unstructured data. Deep learning uses the generative model and Bayesian optimization of evidence just as proposed for life's behavioral models; perhaps, then, techies may help us understand the brain.

Generative modeling of behavior involves circular causality where two dynamical systems are separated by a boundary into *external* and *internal* states. External states are variable and internal states must observe and predict the external states. Evolution of life established the cause-effect relationship. Observation established informational links with beneficial interventional actions for resources. Friston proposes how the best information for the purpose is selected using a principle of least action based on variational free energy. This statistical procedure is formally equivalent to the information bottleneck method which also aids in the artificial information (AI) learning systems of **Deep Neural Networks** (DNN). The network removes extraneous details from the input data by "squeezing them through a bottleneck" and selecting those data most relevant to the problem at hand. Generative modeling treats the information bottleneck as a trade-off between accuracy and complexity that is optimized through minimization of free energy by the network's generative model.

DNN architectures are constructed through an iterative process involving higher layers of the network relevant to the problem learning from the more abstract salient input features. This scheme begins by detecting low-level features of the data that are deemed relevant. These features are then incorporated into an internal model that represents the data in a manner that permits the model to predict what future input data might look like. As the algorithmic programs iteratively extract features from the data, DNNs improve themselves by learning to identify and ignore irrelevant input data and select the relevant. We have treated this as active inference in generative models which represent aspects of an environment for behavioral interaction. Wherever we look at information being transformed into knowledge for a purpose it conforms to Bayesian inference. For behavior interaction, Marr's three levels of problem, algorithm and implementation apply with the process openly obvious in machines and their generative models.

Renormalization Group: theory construction at multiple levels.

We became deep learners through evolution. Deep learning is a broad set of programs that use multi-layered representations of input data to learn which features are relevant and which are not. It is intimately related to field theories. Field theories, introduced in Chapter 5, provide the basis for constructing theories whose component properties may be emergent and need not be fully explained by theories focused on underlying levels. These effective theories find their justification in the theory of **renormalization group** (**RG**). RG is an iterative coarse-graining scheme where fine grained groupings with their properties hierarchically combine with new properties emerging at higher levels. RG as a concept extracts relevant features (structures that represent an operation) from a physical system composed of aggregates of high-energy microelements which is examined at different length scales. As the scale increases, it creates higher-levels with emergent properties requiring their own theories. In Chapter 5 we illustrated this with tiny high-energy bosons and fermions successively aggregating into elements which structure classical physics. This progression has application for hierarchical agency in biology, with information on the data continually merging into greater meaning relevant to the purpose of the problem to be solved.

Renormalization groups in behavior may be considered as hierarchical priors (beliefs at lower levels) with their properties combining into priors or beliefs at larger dimensions where the new emergent properties can be employed in effective theories. The process produces higher-level properties emerging from lower level properties in ascending levels. RG provides the rationale for the use of emergent properties as the basis for their own level of effective theories but recognizing that the generative functions for priors or beliefs apply through all levels. In life the hardware is biochemical and has been selected for a purpose as heritable adaptations. In these living systems the hardware is structured to represent the program for adaptive actions and becomes an algorithm. The biochemistry is enabled and complemented by the information represented in the program so the physics can carry messages. These emergent cognitive properties are functions and we assign names to treat them in theories of behavior. We treated biochemistry in action as cognition in Chapter 3 for chemotaxis and in Chapter 4 for metabolic pathways and the life cycle of a cell as information associated with increasing structured complexity as adaptive behavior. Biochemistry with ion flows also supports nerve nets in the brain with our behaviors that become mysteries in their details.

After trying to cope with the esoteric topic of RG, perhaps I should justify my reasons for this focus. My interest in RG began with Philip Anderson's 1972 Paper in *Science* "More is Different" where he pointed out how in aggregates of matter some properties could not be explained by analysis of their constituents. In other words, new properties emerged that were not predictable. His work contributed to formulation of the renormalization group theory to point out reductionism did not imply constructivism. We have taken a constructivist approach, with emergence of new properties in living systems, to the phenomenon of cognition, building it from the functions that appeared with the evolution of behavior in life. The iterative levels of graining of information fits well with increasing complexity in evolution and the information processing in learning. RG can be thought of as the shedding of information at a lower level as the higher level acquires new properties that apply to theories at that level.

The totality of information involved in any behavior is immense, and for agents to communicate between systems requires that they select a subsample from the total information flow. The meaning that is generated by the agent's generative model provides information on properties of the system the agent must interact with. Predictive success of the messages to other agents establishes the rules of their cognitive language. This requires multi-agent systems to network by short and precise messages between levels of agency as ongoing process is reviewed for its value to other levels. As we go from lower to higher levels, the elimination of less salient information increases timing, efficiency, and predictability. This is the process of coarse graining by which renormalization group theory provides higher-level meaning. Adaptive behavior requires this information bottleneck in order for the generative model to optimize the trade-off between accuracy and tractability. Deep learning is used to build and improve each hierarchical level of an individual's generative model as it interacts with its complex world.

The parsimony of behavioral properties and functions at each level of multi- agent systems make them useful in theory construction, both at the level of the overarching problem and the specific algorithm(s) used to address it. In the last chapter we reviewed two features that provide generality to an effective theory. The first feature from Sean Carroll's *The Big Picture* is, *"For any one effective theory there could be different microscopic theories that give rise to it."* Thus, there is a need for independence from platform, without which we would need to know all the underlying details about how macroscopic behavior emerges from the levels below it. Thus, cognitive psychology of humans, free energy and generative models, agency and behavior throughout life's diversity, control theory, robotics and AI can all treat physics designed for adaptive behavior with a general theory. Deep learning derives from the brain and neural net theory using humans for heuristic purpose. Carroll's second feature was, *"That for any effective theory, the kinds of dynamics it can have are generally limited."* This is borne out by our comparison of the variety of approaches to behavioral studies, all aligning in Chapter 4 on page 11 where we present them with Marr's three levels.

The autobiographical consciousness which we use in discussions of our experience presents the mystery in consciousness. In evolution, was this perhaps just the story of free energy minimization through perception and adaptive action that are considered unconscious processes? These would be stored in memory systems: available for use but not as words. The physics deals with properties that give shared meanings to words. Nouns have properties used to identify entities; verbs have properties that characterize actions; adjectives and adverbs enhance the meaning of each. We have this grammar available in our autobiographical memories as we record them and as we resurrect them from explicit memory. We require the stories associated with experience to guide multiagency-generated homeostasis within individuals and for social communication between individuals. For communication, we supply these stories with introspective properties, which often go beyond the physics and create questions regarding their reality. Perhaps the big mystery is the creation of the culture and the language, which constructs shared stories as vivid metaphors of the physical processes that produced them.

A central role for metaphor.

To effectively share information, one must find common ground with the audience by searching for similarities in functional meaning. Sharing information between behavioral models is a similar problem. The evolution of metaphors as a cognitive sharing of related processes between mental domains was suggested by American cognitive linguist and philosopher George Lakoff (b. 1941) in his 1992 paper, *The Contemporary Theory of Metaphor*. Lakoff's work implies a cognitive basis for metaphor that provides universal utility for communicating functional information.

Friston's theory of hierarchical generative models treats former priors becoming new priors with broader but related meanings. Lakoff proposed metaphor evolution as the cognitive sharing of related process. Cognitive sharing is the basis of communication between multi-agents which begins at the biochemical level, running the life of the cell, and goes all the way up to its expression in the language we use to communicate our stories. All of our priors, inherited and learned, have been shaped through Bayesian selection of evidence in generative models by shedding information for salience. The higher-level priors are metaphorical representations emerging as our beliefs. These, as our beliefs, have lost the history of their micro-determination through the information bottleneck. The final products lack a reductive explanation of their creation and create questions about their reality.

All of these thoughts arise from questions concerning what is real, how life acquired meaning, and what is the difference between living and non-living systems. What we can know about science-based reality is given perspective in a paper by Bricmont and Sokal on the continuing regress of reduction and the reality of new emergent properties as higher levels of analysis are attained. But it is "properties" that we talk about in our search for shared understanding—and just what are properties? *"It is wrong to think that the task of physics is to find out how nature is. Physics concerns what we can say about nature"*...stated Danish physicist Niels Bohr (1885-1962). What we can say about our environment involves two things: the physical structures out there and the data that we can observe to provide information. Our information used as evidence derives from signs receptors convert to symbols. These become attributes with meaning for the purpose of the model. The model is an observer that creates adaptive behavior which in turn interacts back with physics. Our perception of the physics constitutes the evidence the observer must use for behavior. Organisms' sense systems are shaped by evolution to be selective in environmental energies chosen for transduction into sense data for percepts. The total reliance on percepts and what is experienced for environmental knowledge was the position of Baruch Spinoza, Immanuel Kant, William James, and others and seriously treated by Bertrand Russell and the early Ludwig Wittgenstein in their logical atomism. Russell argued that all our knowledge of external things is only inferred from our percepts. He felt that our experience is what we must start from and espoused neutral monism. The latter topic is thoughtfully covered in its variants in Stanford's Encyclopedia of Philosophy Web site, *Plato* (plato.stanford.edu/). Discussing Neutral Monism, they give Russell's position as: *"A basic entity is neutral just in case it is intrinsically neither mental nor physical."* Below is the *Plato* section on consciousness, where they solve the problem by invoking first principles rather than dealing with emergence.

> *"6.2 The Mind-Body Problem. Mach, James, and Russell agree that neutral monism solves the mind-body problem. Russell's account of experience (of perceptual consciousness) may serve as an illustration of the point. Russell frequently emphasized the miracle or mystery involved in traditional accounts of perception. At the end of a purely physical chain of causes there mysteriously arises something of a completely different nature: an experience (a sensation of red, say). This is the hard problem of consciousness.*

> The neutral monist solves this problem by maintaining that "we cannot say that 'matter' is the cause of our sensations". Mach agrees: "Bodies do not produce sensations". To suggest otherwise is to rely on "the monstrous idea of employing atoms to explain psychical processes". Matter/bodies are, after all, nothing but collections of neutral entities, i.e., of Russellian events or Machian elements. **So, mind-body causation reduces to causal relations among events or elements. And, for all we know, the event/element causing a sensation may be quite similar to the sensation it causes**. This closes the apparent chasm between the "material process" and the ensuing experience, and the mystery of perception vanishes."

I rather liked the statement I made bold since I feel that the behavioral interaction is the conscious function of control and editing of an experienced behavior. This goes along with the necessity for an observer and representation for a behavioral model. Sean Carroll informed us (Chapter 5 page 4) that causation appeared with boundary conditions. We considered this as life supplying these conditions for its intervention in physics for a purpose. Events for living things now have a cognitive basis with purpose and meaning. In a generative model, the **event/element** combines both the physics of implementation and the information flow associated with the problem into the autonomous development of the algorithmic program. The information encompasses both the properties and their meanings and the mind-body feeling **reduces to causal relations among events or elements**. The generality of autonomous behavior arises from a combination of action in the external physical environment and a consistent corresponding feel of the flow of information that it generates within. This means that a thermostat is engineered physics while control of our body temperature is a cognitive behavior [because the thermostat doesn't feel anything when it responds]. An interesting paper by Kanai *et al.* (2019) compares one of the generative models of deep learning to brain function pointing out overlapping in cognitive functioning. The sorting of evidence to build the generative model is treated as the generation of information with meaning for that purpose being the function of consciousness.

Observations, representations, or percepts are initial data for a behavioral model, and they are cognitive as information with meaning. Consciousness is emergent, not something that we have to deduce from facts of physics. As initial data, percepts are the bedrock of cognitive properties. It is therefore a mistake to try to explain the existence of cognitive characteristics (or qualia) as emerging from complex brain states since these characteristics emerge from percepts, at the very beginning of behavior when molecules became both physical structures as well as perceived information. If we assume cognition only appeared as a brain function, we quickly flounder in reductive explanation. We present behavior as a cognitive function that goes back to life's beginnings.

This perspective provides a basis for the concept that metaphorical similarities in priors generate different but similar responses at different levels of complexity. This creates informationally similar systems that may serve different purposes and may be implemented on different platforms. This concept pertains to any self-constructed autonomous system with adaptive heredity that must operate in a complex variable environment in order to survive. These systems must continually interact with and intervene on their environment in order to obtain the energy and resources necessary to maintain their existence. Life is continually faced with problems it must solve. Before life, problems did not exist, and matter and energy followed purely physical rules established at the beginning of the universe.

The discussion of neutral monism helps me think as a monist, believing everything is based on just the stuff that originated in the Big Bang. However, I am also an emergentist that believes properties of aggregates of that stuff both can differ from those of its constituents and not be predictable from them. Reductionism does not supply all the information for constructivism. It was the emergence of life and its informational properties that introduced purpose and meaning to aggregates of stuff (matter, energy, information, and properties). This transition made stuff more interesting, rather than just being material following rules. The mystery of consciousness lies in information gleaned from the environment and the meaning imparted to that information.

The nature of "properties."

Properties are what we talk about and properties of neural nets are what do the talking. To be rigorous we must deal with two sets of properties. These comprise, firstly, the properties of the observer, where internal information represents aspects of the targeted external system; and, secondly, those aspects of the targeted system that the observer models and interacts with. This all takes us running in a circle where percepts are used to infer properties and causation in our physical world, and then those inferred properties themselves are used as evidence that our physical world exists as we perceive. This duality happened when we developed the complexity of communication using our language, nearly four billion years after the evolution of behavior began. It is living systems that perform all the analyses that are required to establish the two sets of properties that integrate knowledge of adaptive physical causality into behaviors. Formation of the behavioral models that drive agency depends on the ability to construct percepts of the world. This formation of models is where information, knowledge, value, purpose, and percepts themselves appeared in adaptive behavioral systems.

The simplicity of these beginnings, acted on by natural selection, allowed those with a reductionist perspective to claim that life was just physics and luck. It is a story we created to justify using causal constructivism without cognitive control properties of generative behavioral models. A similar stance could be employed toward the evolution of behaving neural systems that didn't require a higher nervous system like our brains. Once our complex nervous system did evolve, it provided new dimensions to our percepts. These dimensions were considered to be beyond physical explanation, and while their properties must have emerged somewhere during the evolution of animals, it was not clear when and where that was. We find a similar mystery in the world of physics where the properties of the classical physical world overlie and ultimately depend on the non-intuitive subatomic world of quantum mechanics and wave functions. Our introduction of effective field theory and renormalization group was introduced as physicists struggled to account for emergent properties. Properties that were assigned to larger-scale classical physics were viewed as real, intrinsic, and inherent to the inferred physical world of our creation. We cited the Bricmont and Sokal paper on reality in science that treats their *"renormalization-group view of the world"* and there is a thoughtful quote from their summary that applies here on the shedding of information in the coarse graining of ascending levels.

> *"The theory on each scale emerges from the theory on the next-finer scale by ignoring some of the (irrelevant) details of the latter. And the ontology of the theory on each scale – in particular, its "unobservable" theoretical entities – can be understood, at least in principle, as arising from the "collective"*

or "emergent" effects of a more fundamental theory at a finer scale."

The discovery of a hierarchically layed physics occurred after Darwin and his theory of evolution, which introduced the concept that living systems arose from the physical world by natural selection for a purpose. Experience can be viewed as practice and it guides selection by inference in generative models of adaptive behavior. Data gathered via our senses were used as evidence to both build our behavior and to describe the properties of classical physical systems. Thus, our description of the world is a combination of both the physics of the behaving system and the flow of information in our models and the physics of the target. Percepts of properties can also be viewed as emergents from classical physical systems since the cognitive processes that infer their existence emerge from representation of those properties in the model. Science can provide observations of the physical structures and processes of the behavior but cannot measure how the flow of information feels to the model as an experience occurs. Life uses its cognitive model to first form representations of sensory data, then converts these representations into information that possesses recognizable properties or qualities. The intrinsic meaning and value of this information means the physical system from which it is derived can be shared, but we must create stories about how the corresponding information feels.

Cognition must be explained by theories that encompass its functional properties: percepts, information, knowledge, meaning, purpose, and value. These properties provide the functional basis of the cognitive aspects of the algorithm's hardware of a living behavioral model. The functions of the properties are general in all models, but the contingencies associated with separate derivations of the same or similar behaviors render variability. The variability is a function of heritability in a lineage, environmental influences, degree of individual learning, and complexity. To tell stories about the physical world we use external examples that we can point to that all will agree are using the same rules. This has also become the role of mathematical models in the extension of physics to treat unobservables. One of my favorite Bertrand Russell quotes reflects his commitment to percepts as our evidence:

"Physics is mathematical, not because we know so much about the physical world, but because we know so little: it is only

its mathematical properties that we can discover." Bertrand Russell (1927). *An Outline of Philosophy*

Physics follows stable rules while living systems are based on unique paths with contingencies and counterfactuals requiring individual arbitration. Living systems that can communicate make such arbitration possible, permitting them to reach agreement on how to best adjust rules in order to play the game well. A substantial part of 20th century philosophy focused on language and its uses—good and bad—in communication. Notably, Ludwig Wittgenstein in his later years moved from his logical position on the analysis of meaning in language. He thought deeply as he tested the concept against all the different ways language was actually used. These deliberations led him to his theory of '**language games**'.

Wittgenstein and language games.

Wittgenstein developed his notion of language games in his treatise published posthumously in 1953, *Philosophical Investigations*. To provide a simple example of what language is and does, he wrote: *". . language is meant to serve for communication between a builder A and an assistant B. A is building with building-stones: there are blocks, pillars, slabs and beams. B has to pass the stones, in the order in which A needs them. For this purpose, they use a language consisting of the words "block", "pillar" "slab", "beam". A calls them out; B brings the stone which he has learnt to bring at such-and-such a call. Conceive this as a complete primitive language."* The nature of language games is language emerging as a function in social behavior. Thus, the concept has application to achieve the purpose of multi-agent behaviors. We can conceive of this as the conversation involved in interactions and compromises that generate homeostasis in the cell.

In 2012 the computational biologist Jonathan Karr and collaborators published a "whole-cell" model of the bacterium *Mycoplasma genitalium*. Since the real cell is too complex for real time analysis, they simplified it into a simulation. The simulation used a series of modules that mimic cell functions, representing the interactions of 28 categories of molecules including DNA, RNA, and proteins and their substrates and products. The cell functions organized these molecules into the above modules, which communicate with one another by exchange of data and instructions back and forth at one-second intervals. In living cells, this communication between cooperative modules leads to complex behaviors. Their 'language' represents primitive symbolism, with molecules carrying the messages. The biochemical pathways conform to 'builders' requiring substrates as materials from their 'assistant' that supply the required substrate as the information chain linking their behaviors.

As Wittgenstein grew older, he became more pragmatic and holistic as he considered language to be communication of concepts associated with life's activities. This perspective overlaps with George Lakoff's ideas about the importance of metaphors in communication by language. Metaphors achieve meaning from the similarities in functional roles that are shared by the recipients of the message. Functional overlap works at the level of the generative model, which communicates internally and externally. Early in the chapter, metaphors were discussed as priors with symbolic similarities. Facilitated pathways responding to similarities in input carry this to nerve nets, introducing metaphors in speech. This is the implication in the last sentence of the epigraph by George Lakoff used for Chapter 7, Life is a Story: "…. *the locus of metaphor is not in language at all, but in the way we conceptualize one mental domain in terms of another.*" Linking behavioral priors into more complex behaviors is facilitated by their metaphorical similarities. The linkage of models produced the metabolic pathways for the cell treated in the above paragraph. Recognition of the metaphorical similarity within and between priors was one of the emergent cognitive functions not only for communication between models but also playing a role in resource recognition. If something like language is required for communication, it must employ words or symbols.

Wittgenstein's "use theory of meaning" says words are not defined by reference to the objects they designate, nor by the mental representations one might associate with them, but by *how they are used*. Sensory data are transduced to information as symbols incorporated into the model as a function. The role of free energy is to optimize predictability of the symbols and their functional role. In the previous paragraph we characterized homeostasis as a generative model that had minimized its free energy through adaptive communication. This process requires a network of adaptive prioritized information processing with updating and control free of error and bias. All of this communication must be based on a symbolic language that allows information to flow through the behavioral models. To begin this chapter, we introduced comparisons between computer deep learning and the generative models of behavior in living systems. As these systems became complex and multilayered, the information processing treating them in computers became heuristic models for living systems.

Evolved behavioral networks must shed massive amounts of input data in their construction and use. Extraneous details and those less salient are eliminated by squeezing them through a bottleneck and selecting those relevant to the concept. Friston treats the information bottleneck as optimizing the trade-off between accuracy and complexity in the generative model. This is the role of Bayesian inference used to optimize the behavior of a model. As experience is repeated and compared with diversity of similar experience it becomes a probability distribution that narrows to a salient peak. Functional salience is metaphorically responsive in recognition of variability and its refinement. Friston's work involved with generative models of living systems could be considered as learning the rules of the language game treating communication of symbolic information with meaning. 'Words' become the symbols dealing with metaphorical similarities as probability distributions seeking adaptive solutions to problems.

Our brains possess a hierarchy of internal models of aspects of our environment that we use to predict sensory data, and to explain these data in terms of their causes. With new input, we update our beliefs to form a posterior belief about the world. This requires the passing of local messages between variable and successive levels in the decision-making generative model. The passing of information from lower-level models of features and properties at micro-scales leads to selection of relevant higher-level features based on a probability distribution of previously-experienced outcomes. This weeding of information for relevance is the process of generative models. Recursive passes with error correction by experience optimize the models for simplicity and accuracy and updates the model as conditions change. These models arose nearly four billion years ago and some of the genes still run our life processes. Life has a long history for framing 'language' that is shared, stable, and heritable to effectively communicate. Using information is how agents communicate to build their generative models and is also the basis of functional language games.

Consciousness, observation, percepts, and generative models.

Here, consciousness is considered as overlapping with observation, representation, and perception and thus as a general cognitive property, and we explore its functional role in the generative models of behavior. Consciousness plays a role in the generation of percepts from sensory input. To handle variability, representation of percepts will be a probabilistic model optimized by selective inference. The representation can be viewed as the metaphorical match between the environmental target and its use in the model, where the feel of the match reflects prediction of salience. In the generative model itself, this is the functional outcome of minimizing the variational free energy for salience when models communicate. The functional role for consciousness is generating the feeling of how well the output from the generative model matches the desired behavioral purpose. Those that are linked in networks of biochemistry operate autonomously below our awareness. Those dedicated to poetry and music achieve deep feeling as it is experienced. We have accepted Thomas Nagel's position that consciousness feels like something. In Friston's models we apply this feel to salience and similarity of information on symbols for selection to priors. This would fit the free energy function in active inference. Similarities between behavioral domains create feeling for Lakoff's metaphors. Metaphors that feel right create communication relating to information becoming primitive language. For the rules in different language games, Wittgenstein uses the correct feel of functional metaphors. Consciousness as a feeling is inherent to behavioral models and has value in predicting if our models will work as our world presents challenges and opportunities. Consciousness edits behavior and plays the key role in feeling good or bad about its performance, which emphasizes that consciousness should be Bayesian to be rational. Some recent studies provide help in unraveling how and where attention is focused on detection and correction of errors in performance of our behavioral models and how personal experience plays a role in this.

Unguided versus guided attention: the DMN.

The American neurologist Marcus Raichle (b. 1937) used brain scans to image brain areas that showed organized activity when study subjects in the scanner let their minds wander. The activity in these areas then decreased from that baseline when they concentrated on a task. This effect was initially associated with mind wandering and daydreaming where there was no goal or purpose for engaging the mental machinery. This was labeled as the **default mode network (DMN)** of broadly connected brain regions, which are more active when we are not engaged in a focused task. When our consciously focused attention shuts down, the default mode network broadly activates as if it were the dark energy of the brain; metaphorically comparing 68% of our Universe as the hidden dark energy accelerating its expansion. What does the brain do unconsciously when released from the conscious control? Why develop a brain network to spend energy goofing off? Perhaps the activity when mind wandering was associated with the high level accomplishments of our conscious mind. I became interested in the functions associated with activity of this highly connected network of the brain. The work on the Bayesian brain described the formation of models to predict the best response to input through the optimization of generative models. Our functionality at getting it right is amazing. How and where did all this work get done?

The function of the default mode network intrigued the interest of many scientists who extended the research of the DMN. Their work reveals that the DMN is responsible for a diverse set of activities that were previously attributed to consciousness. The following list of suggested functions associated with DMN activity suggests how this highly interconnected multimodal network might help create our mind's versatility.

- Autobiographical memory
- Prospection to imagine a future
- Generating cognitive scenes and episodes with context
- Recognizing current external context for response
- **Theory-of-mind** reasoning for social interaction
- Social understanding: emotion, empathy, manners, morality
- Attention focus both active and background
- Status evaluation for predictive processing
- Autopilot to employ well practiced behavioral models
- Ongoing optimization of salience and relations of models
- New ideas from diverse knowledge and themes
- Recognition and attentional focus on these ideas

Regions of the brain associated with DMN are only sparsely connected at early ages and through ontogeny integrate into an interconnected network. Without a properly functional DMN, we lack some skills necessary to coordinate engagement with the external world. DMN dysfunction has been related to a broad spectrum of problems: dementia, schizophrenia, epilepsy, anxiety and depression, autism, and attention deficit/hyperactivity disorder. A well-functioning DMN has been related to creativity in thinking without the use of words and using conceptual representations. Einstein told us, "*Words or the language, as they are written or spoken, do not seem to play any role in my mechanism of thought.*" Creativity is neither logical nor explicit but involves the ability to shift between different modes and themes and recognizing metaphorical priors that lead to a novel hypothesis. We think in neural nets and then put thoughts into language to share them.

Our consciousness builds our autobiography.

Why do humans have their dramatic experiential consciousness that they think runs everything, when the richness of processing and decision-making lies hidden in our subconscious? Our consciousness is comprised of selected fragments run through the internal narrative that we build from our experience as we participate in the world. Did our consciousness specialize as an adaptation for social and cultural communication when language evolved? We all need to point to features and communicate in order to agree on meaning for the shared symbols represented by words and syntax. This agreement is necessary in order to translate the beliefs generated by our memories into shared socially-agreed upon metaphorical communication. The properties of our default mode network, our autobiographical memories, and conscious awareness provide a trail of correlative interactions for this social function.

This process also creates an ongoing cultural story of shared experiences that we can then replay in our autobiographical memory. In the richness and diversity of our world's cultures and meanings, how do we sort out the story we want to participate in during our lives? Experiences of most species are strongly shared through similarities in habitat and behavior, while our experiences are massively polymorphic. This became increasingly important as our species moved from hunting and gathering in small groups into the acculturation that led to our complex and diverse civilization. Coincident with this geologically short history is the explosion of within-individual diversity in experienced lives. Learning a world-view is essential to an individual and is its life work. An individual's personal history of social interactions as it learns to fit into their world creates unique stories. The need to mingle our individual stories with an exponentially more complex world differentiates us from our near ancestors and all other species. From a personal perspective, we can ask: what is important; how does it work; what does it mean; where is it headed?

Our consciousness participates in the creation of our DMN diversity and accesses it to facilitate our interactions with others.

Socially adaptive learning requires arbitration in how the DMN functions listed above are performed. Societies suffer when the stories of some groups *unjustly* blockade the stories that others wish to tell. The integration of diverse themes is the challenge for today's societies. This challenge must be confronted on the level of individuals but requires adjustments in their goals and values. Social integration begins to be shaped by early life experiences. Our attitudes and values later respond to peer groups and charismatic leaders who often have strong appeal to a group's shared social values. Like most things in life that are built through interactions, the DMN and the conscious are recursive and adaptation must use new evidence. The more open we are to the use of all the evidence, the more adaptive we become in our ability to predict the future through Bayesian inference.

Unguided DMN activity sheds information, and when conscious attention isn't focused, it directs processing to increase salience of our stored memories and their interrelations. This involves selection of the best evidence and it is up to us to decide what to believe, its probability, and what evidence to input. Without conscious attention extraneous details are shed, saving those relevant to the concepts of behavioral priors. This optimizes the trade-off between accuracy and complexity for the models of our experience. When unexpected surprising input is encountered, conscious attention is evoked to resolve our response. The new evidence can either be dismissed or lead to prefrontal processing. Fast decisions are more subject to the bias of long-held beliefs and new information may be ignored or discounted. When we are surprised, we should question beliefs and search for evidence for model updating.

To understand consciousness as a function, the generative modeling of computers provides a helpful model. As I reflect more on the computer screen and the screen in our brain, it becomes apparent that neither computer nor brain needs the display for routine work. Each model has its own representation contained in information flow in the generative model. A quote from the Introduction to a paper, *"Latent Variable Modeling for Generative Concept Representations and Deep Generative Models"* by Daniel T. Chang provides a model for a hidden consciousness that functions for information processing.

> *"Latent representations are the essence of deep generative models and determine their usefulness and power. For latent representations to be useful as generative concept representations,*

> *their latent space must support latent space interpolation, attribute vectors and concept vectors..."*

Elements from these working representations are selected and made available when interactions with other models are necessary. These operate in latent space and are thus hidden in their representation. In the computer, selection of processing relevant to the machine operator can be provided on a screen or voice. These are a re-representation on the screen for interactions with the operators. The goal for the machine is to share its behavior with its builder in a form the operator can use to interact with the machine's latent representations.

Our conscious virtual display is called up by a high level of attention to prioritize interactions with input. Our DMN can access our autobiographical memories to interact with new decisions. This virtual display we relate to as our conscious awareness is a high level meta-consciousness that we enhance with language and stories. The DMN produces and interacts with this re-representation. I feel this is the increasingly important role played by the consciousness as the critic, editor, and motivator of the choice of input. Our hidden processing both produces the attentional focus and selects input from it for processing and potential refocus of attention. How can our consciousness influence the background processing that produces it? We must change the path for consciousness in creating ongoing attentional focus by changes in the background that select it for our display. Our interests and values have biased our attentional focus for predicting the role an input may play in our future. Openness in questioning creates enhanced input for potential interactions with beliefs. If motivated, we begin model modification to create better predictions from new evidence. This remodeling effort depends on the effectiveness of our editor during our lifetime in creating adaptive knowledge, interests, and values.

This process can serve as our mentor and therapist. Consciousness must act as a guide in the development of a functionally adaptive DMN, especially since the call to attention is evoked by the DMN. There is certainly circularity here since recursion is the basis of generative models. Since attention is rooted in our limbic system of emotions and values and associated with our autobiographical story, change is not easy. Rapid change may result from joy, trauma, or cognitive dissonance, which leads to highly motivated conversions while revaluation of reward systems proceeds more slowly. There are personal stories of experienced contingencies that begin reshaping behavior to create paths to different futures—and this book is my story.

The DMN is a positive feedback system that, as it becomes more open and diverse, can better interact to guide adaptive development. Babies and toddlers are the most open and rapid learners, driven by attention to what is new and questioning what it means. The opposite direction is also possible and appears to dominate in society. A personal history of being open or closed to new input sets the pattern of what gets in. The ability to assess the value, reality, and importance of evidence is crucial for functionality. Individuals program their DMN's attention to select what it should work on. This stresses the importance of considering what could be (i.e. counterfactuals) in order work toward a better future. Promoting a better future depends on maintaining surprise, wonder, curiosity, questioning, and motivation. Our ability to be flexible and adaptive depends on the diversity and relevance of evidence that we select to adapt to change. We are constantly involved in predicting and deciding the best way to respond to the next input. For enhanced performance, this involves the use of episodic memories from our memory store to be parts of new models. We must constantly form an appropriate model to respond to new input. This activity entails using experience from our past as beginning hypotheses. The important role of consciousness is how we choose and cross-file our store of episodic memories.

The gist of it....

When asked how we would know when computers had reached the intelligence of humans, the British anthropologist and linguist Gregory Bateson (1904-1980) answered, *"We will ask it a question and it will respond, 'That reminds me of a story.'"* If the question fits a story that the computer or we know we have the answer; if it is new but evokes stories, we know it provides direction toward the answer. We begin to ask questions about the stories we know that share concepts to see if we can creatively adapt old stories to the new question. These stories are filed and indexed as pieces of memories in our brains, and to access the similarities requires indexing to make them available in a search We have all had the experience described as a sense of knowing, tip of the tongue or déjà vu. We treat this as involving the **gist** of a closely related memory. The first glimpse of scene while flipping channels on television provides meaning at levels ranging from perceptual to conceptual. The gist of that meaning can be accessed for greater depth for different purposes. When attention calls us to explore a problem, it is the gist of similar stored episodes that enable us to search of our memories.

Frames (a scene from a story), scripts (action sequence of a story), themes (similar stories) and case based reasoning (adjust a similar case for better fit to the current situation) are aids for processing these stories which came from Roger Shank, Marvin Minsky and others in AI as tools for learning to operate in a complex changing world. These tools emphasize the importance of diversity, depth, and cross-referencing by degrees of similarity as we process stories of our behavior. This is how we store and organize our knowledge of the environment in which we operate. We recognize features of percepts of components of the input that are similar to previous experiences. An interesting concept introduced by Shank was an indexing scheme for memories that facilitates their acquisition for adaptive reuse. His theory of MOPs and TOPs (memory-organization packets and theme-organization packets, respectively) concerns how human memory is organized. From his www.Edge.org bio page *".... any experience you have in life is organized by some kind of conceptual index that's a characterization of the important points of the experience."* This is what we experience as our autobiographical memory, where episodes from our life experiences are recalled in response to focusing our attention on a current pending decision.

Traces of episodic autobiographical memory are distributed through the default mode network, with the hippocampus serving as the hub. Emerging evidence from studies of the DMN related to autobiographical memory suggests that experiences are processed along a gradient of abstraction. This allows the same remembered event to be recalled either as **concepts** of their meaning or **percepts** on the details of the setting. Studies on the storage of episodic memories in a paper by Sheldon *et al* (cited below) reveal that concepts and percepts are separated and sent to different brain regions. The study shows that the anterior hippocampus extracts conceptual and schematic elements as more holistic aspects of an event, which give it meaning. For storage of these elements, the anterior hippocampus is connected to the frontal regions of the brain where evaluative, schematic, and self-meaning processing occurs. This is the part of the episodic memory that has knowledge, value, and emotion and relates to us as individuals. The gist of this event is stored in the anterior hippocampus with links to the percept details needed for recall from their separated storage by the posterior hippocampus. The posterior hippocampus is connected to posterior brain structures that store the context and perceptual episodic representations of the event. This provides detailed images of past experience, which can be recalled when making decisions pertaining to well-structured problems previously dealt with. The anterior hippocampus uses the stored gist of the conceptual memory to access detailed perceptual information that will be used in decision-making. Storing different elements of a given memory in different locations raises the potential for errors when that memory is recalled and reassembled. This leads not only to errors in assembly but also to the imaginative enhancement of memories by the addition of counterfactuals. The association of counterfactuals with real memories enhances a subsequent faulty belief in their accuracy.

Decision problems can be open-ended, and access similar concepts stored in the frontal brain. Or they can be close-ended tasks that have a previously-experienced path that only requires control and some minor tweaking for resolution. Although certain situations benefit from representing our past as either concepts or percepts, it is the ability to adaptively shift between these that provides the power of our autobiographical memory. The brain's role in coping with surprises emphasizes the importance of storing memories as both gist and detailed perceptions, with gist being most valuable when surprises occur. So, memories not only help us recall our past, but help us make predictions for the future. Gist appears as the rapid result of attention, in less than a tenth of a second, to allow us to respond quickly to a surprise. The gist is the beginning of a behavioral model that can then access details from the environment and past experience to assemble a response. This stresses the importance of diversity and continuity of learning, since what we have learned is the ultimate driver of what we will learn as we continue to expand our experience. As our world becomes more complex and dynamic, lifelong learning becomes more critical.

In trying not to violate rules of physical science to address the mystery of consciousness, I became engaged with the perspectives of AI, robotics, cybernetics, control theory, effective field theory, renormalization groups, and philosophy and attempted to merge these with the perspective of biology. When all of this is incorporated into a general theory, it appears that the emergent properties of living systems are intricately connected to the communication of information related to value and purpose. We accept the physics of sending a man to the moon since we see and describe how engineering acts. There is the hidden story of why, where, when, and how the biology made it happen. This story is the experience of billions of years of agents communicating between their programs, shedding unimportant data, and saving more valuable information. We treat behavioral models' representations of symbols used in adaptive control as basic consciousness. Consciousness is not a thing but becomes a function in a process and relates to providing data with meaning to become information. In *E. coli*, chemotaxis as adaptive purpose is complex, but not a mystery since we follow biochemical pathways rather than treating complexities of neural cognition. In deep learning, machines create programs for adaptive purpose that may not be decipherable by programmers. Analogously, we should not expect to unravel human consciousness by measurements on the physics of neural pathways.

I am sure that reading through *E. coli*'s chemotaxis and metabolic pathways did not offer an overall intuitive feeling for what perception or consciousness is or feels like for a bacterium. In gradually increasing degrees, that is true for other people, babies, chimps, dogs, mice and on down the scale of nervous systems to cnidarians. The cognitive control architecture of chemotaxis provides *E. coli* with a perceptual world of perhaps a few dozen chemical and physical entities along with their magnitude and meaning so *E. coli* can decide whether to flop or swim. A sophisticated wine taster may have five or ten times that number of sensed descriptions of taste, aromas, and feels that enables them to extoll, critique and make decisions about where they are in a wine environment. The clever border collie, Chaser, learned 1022 different toys by name and could retrieve each by its learned name. An African grey parrot, Alex, could recognize 50 objects, 7 colors, 5 shapes, count up to 6 and respond to questions concerning combinations of these. Alex had a vocabulary of over 100 words and could communicate at the level of a two year old. Looking in a mirror Alex asked, *"What color?"* and learned that he was grey. He was temperamental and interactive and his ability to speak created something I could recognize as consciousness in the same way we recognize it in children. When his friend and collaborator of thirty years, Irene Pepperberg, had to leave him overnight at the vets for an aspergillus infection she said *"Goodbye, Alex."* Alex was apprehensive and said, *"I'm sorry. Come here. Wanna go back."* Consciousness can take many forms but has the fundamental relation to creating information with meaning and purpose from matter and energy—and we don't really know what they are either.

We have taken a natural history approach to understanding consciousness. We observed that an *E. coli* has very different properties from its constituent molecules. They are alive and we can ask, "What does it feel like to be alive?" The use of some of these molecules in models as messages to produce overt behavior was another level of complexity. They had models used for behavior.

Life and behavior had emergent cognitive properties. Cognitive properties are beyond our basic percepts of the physical world where we have a shared reference through language. Are cognitive properties real is a question that persists? We don't know what energy really is but have an agreed upon real world of physics through shared observations and predictability of concepts from observations we can make. Both systems must use the same laws, but teleology, contingency, selection, and heredity create life's diversity through their use. Using shared experience, we create language to communicate our experience as messages. Only dedicated experts can achieve understanding of complex systems where untold numbers of micro events produce macro events, and this understanding often is only possible through mathematical models. Models and mathematics can be conceptually treated as metaphors. For most folks, sophisticated reductive approaches to explanations are neither feasible nor rewarding. Most must gain appreciation of complex systems by extracting stories metaphorically, treating these systems in prose and pictures. A more holistic narrative theme can be developed in story form, which provides perspective on the system. Since the behavioral models of living systems create behavioral experiences with causality, purpose, value, and meaning, these readily translate into story form.

Further Reading and References.

Callard, F. and D. S. Margulies (2014). What we talk about when we talk about the default mode network. *Frontiers Human Neuroscience.* 8(619). https://doi.org/10.3389/fnhum.2014.00619

Chang, D. T. (2018). Latent Variable Modeling for Generative Concept Representations and Deep Generative Models (arXiv:1812.11856v1 [**cs.LG**])

Kanai, R. *et al.* (2019). Information generation as a functional basis of consciousness. *Neuroscience of Consciousness*, 5(1) niz016

Lakoff, G. (1992). *The Contemporary Theory of Metaphor.* in Ortony, A. (ed.) *Metaphor and Thought* (2nd edition), Cambridge: Cambridge University Press.

Sheldon, S., C. Fenerci and L. Gurguryan. (2019). A Neurocognitive perspective on the forms and functions of autobiographical memory retrieval. *Frontiers in Systems Neuroscience,* 13(4). https://doi.org/10.3389/fnsys.2019.00004

Sokal A., and J. Bricmont (2004). Defense of a Modest Scientific Realism. In: Carrier M., Roggenhofer J., Küppers G., Blanchard P. (eds) *Knowledge and the World: Challenges Beyond the Science Wars.* The Frontiers Collection. Springer, Berlin, Heidelberg. https://doi.org/10.1007/978-3-662-08129-7_2

CHAPTER SEVEN

Life is a story: Poetics, Metaphor, Bayes, History of Models, and Reflections on the Ontology of Concepts.

"[T]he greatest thing by far is to be a master of metaphor… [it is] a sign of genius, since a good metaphor implies an intuitive perception of the similarity in dissimilars." Aristotle. Poetics 335 BC

"As a cognitive scientist and a linguist, one asks: What are the generalizations governing the linguistic expressions referred to classically as poetic metaphors? When this question is answered rigorously, the classical theory turns out to be false. The generalizations governing poetic metaphorical expressions are not in language, but in thought: They are general mappings across conceptual domains. Moreover, these general principles, which take the form of conceptual mappings, apply not just to novel poetic expressions, but to much of ordinary everyday language. In short, the locus of metaphor is not in language at all, but in the way we conceptualize one mental domain in terms of another." George Lakoff

"The Universe is made of stories, not of atoms." Muriel Rukeyser The Speed of Darkness, 1968

In the preceding chapters we have extensively discussed priors of generative models as hypotheses or beliefs about interactive behavior. Hypotheses connect concepts by specifying the expected relationships between propositions which are statements about ideas. We introduced the communication between models as a form of language and examined similarities of models as metaphorical and related to sharing similar concepts. Metaphors play an important role in communication as a language. Metaphors relate to functional roles in the sense of Wittgenstein's use theory of language treated in Chapter 6. When hypotheses function together their information becomes a type of conceptual framework. This is an abstracted representation that organizes ideas into beliefs as high level concepts. When a conceptual framework is complex and incorporates causality as an explanation of purpose it is generally referred to as a theory. Behavioral conceptual frameworks capture something real and do this in a way that is easy to remember and apply. We have made the point that for a broad audience this is best communicated as a story. Let us see if we can tell stories about behavior to achieve another level of understanding.

The first epigraph is from *Poetics*, where Aristotle explains how to treat the drama of life presented as a story. Aristotle recognized that knowing a variety of stories on different themes enabled metaphorical cross-referencing that broadened audience participation. The second, which treats language as metaphors of cognitive process, is from George Lakoff's 1992 paper, *The Contemporary Theory of Metaphor*. Since knowledge and thoughts are cognitive processes, metaphorical analysis has general application to behavior since our generative models can be seen as metaphors of their environment. Lakoff's referral to metaphors as comparative similarities of cognitive properties between mental domains must relate to the sharing of information between models. Metaphors don't only act in adjusting priors in Bayesian inferential linkage between domains in individuals, but also metaphors would work socially in communication between individuals.

The third epigraph relates the Universe story to metaphoric similarities between domains on many scales. Leonardo da Vinci, Renaissance polymath master of many disciplines, informed us, *"Learn to see. Realize that everything connects to everything else."* For the atom, the stories would treat their formation from quarks and electrons, just after the Big Bang. The story of the Universe would relate how these atoms structured the cosmos and our bodies over fourteen billion years. This provides a rational basis for Sean Carroll's poetic naturalism as a language to treat emergent properties in an effective field theory for both physics and cognitive studies. Metaphors represent thematic overlap of informational similarities that provide context and content of emergent properties. They can be words, patterns, symbols, signs, or icons. Metaphors represent properties assigned to entities for theories in science to treat levels of emergence from complex micro-processes. Mathematics treats these properties in a formal way. Human history with communication for behavior built a bridge linking the physical environment with a dramatic story sharable with spoken language. Life process can be told as a story rather than formal physics or chemistry and reach a much broader audience. Life really only makes sense through the story of its developmental history and the meanings associated with paths chosen. Perhaps Aristotle perceived the relation between metaphor, information, and communication of meaning that is the basis of cognition; let's explore this relationship.

How to build different paths in life's story.

The world presents many choices and we choose paths leading through this kaleidoscopic environment. Our ongoing neural activity selects novel items that enter the consciousness of our working memory where we can incorporate these items into our existing life story. We compare these new items with our constructed story for validity and value in order to edit the story. As we continue to rewrite our story it also directs us in which paths to choose. The story we tell is written scene by scene as our view of our recent experiences and emphasizes aspects to which we gave the most importance. Ideally, we should compare this story to its success in predicting choices that affect our future wellbeing. Our ability to predict will in large part depend on how well we know our environment and its diversity, probabilities, and causal links. The plot of our story should be edited using the best evidence, not just our emotional response to new elements. To be rational, our acceptance or rejection of evidence must be based on the real world, not just our prior belief in our edited story. Surprising observations should supply motivation to edit or rewrite our stories to tell them better.

Building different paths in neural and biochemical stories.

The availability of useful relevant knowledge is also the rationale for Jakob von Uexkull's concept of Umwelt described in, *A stroll through the Umwelten of animals and humans* (1934). He worked on the fit of a species' sensory abilities to its behavior over the diversity of species from amoeba throughout the Kingdom of animals. He made the case that species experience life as specific to their sensory abilities used to model their subjective environment that he called *Umwelt*. The psychologist, James Gibson treated **affordances** in *The Ecological Approach to Visual Perception* 1979 and defined an **affordance** as behavioral uses the environment furnishes the animal related to their behavioral ecology. Both of these concepts are equivalent to Bayesian priors that have evolved through evolution and learning by preceding generative models. They become accessible as priors that are heritable or as an individual's memories. These are cognitive elements to be selected by **abduction**, which is getting the right stuff at the right time from our memories. This provides our best guess as to how we should interact with an input for the same or a metaphorically similar problem that we encountered in the past. The memories we select to generate our behavioral response are those that conform most closely to the current situation. These establish a new prior that is then refined by Bayesian updating with continued experience and inference. This process can be viewed as metaphorical in nature because we respond to a new problem by first comparing it to our memories of apparently similar events in the past.

In neural systems, similarities in the input of the new problem evoke memories that become parts of the new emerging model. Early sensory neurobiologists suggested the nervous system detects rather than projects specific features of a stimulus to assemble perception. During the 1950s innovative neurophysiological investigations confirmed this scenario in frog and cat vision. Later this view was extended to fish that sense and detect objects with weak electric fields and bats that use sonar to do the same. Worlds are constructed and shaped from the neural priors we employ to predict the input. How are the similarities selected? In neural circuits we simply rely on 'neurons that fire together, wire together' with those selected from priors as previously facilitated paths. These neurons would recruit existing neural networks through similarities in input evoking parts of synaptically wired nets that match closely enough to respond. The more familiar the input is, the better the match. Recurrent passes as the net grows expand and enhance recruitment. Thus, assembly of a hypothesis begins with the history of the generative models. The hypothesis is shaped to fit the new use by active inference selecting inputs to optimize the model. For learned behaviors, after many repetitions, connectivity of these specific neural paths has high probabilities. They are hard wired. Some neural circuits for behaviors come prewired as programmed behavior with developmental and evolutionary origins. Hard wiring essentially increases probabilities of the appropriate network firing, but for their use they employ recursive error correction and controls.

Biological systems continually diversify in directions that are not predictable and the potential for diversity is a function of both the organism and its environment. Early in the evolution of life priors, representing properties of a resource, became the start of heritable biochemical agency. As a model replicated it could grow in size and become variable. Variants can provide the potential of new priors for other adaptive interactions. Previous traits assuming new uses is the process of preadaptation proposed by Darwin in *On the Origin of Species*. Change in function can begin to develop new uses in structures and behaviors evolved for other reasons. To stomp out hints of teleology, Stephen Gould proposed a name change to exaptation and produced an expansion of interest and papers. In his evolutionary work, Stuart Kauffman came up with the concept of the **adjacent possible** as the unpredictable ways in which the present can reinvent itself for new purposes. Steven Johnson popularized the concept in his book *"Where Good Ideas Come From"* to describe how knowledge in one area may provide a head start for solutions in a new problem area. This was presented as a formal theory, *A Quantum Model of Exaptation: Incorporating Potentiality into Evolutionary Theory*, by Liane Gabora, Eric O. Scott and Stuart Kauffman that proposes a framework treating key features of exaptation that both enable and constrain potential future changes. They propose adaptations in evolution, learning, and culture both facilitate and inhibit future directions.

> *"A model of exaptation must incorporate the notion of potentiality: every biological change not only has direct implications for fitness and so forth, but it both enables and constrains potential future changes. The notion of potentiality incorporates both the 'adjacent possible' those states that are directly achievable given a certain initial state, and the 'nonadjacent possible', those states that are remotely achievable given a certain initial state. Exaptation occurs when selective pressure causes this potentiality to be exploited."*

Priors as metaphors are derived from the biochemistry of life as well as neural elements. Products of biochemical evolution were fashioned into the ancestral cell. The cell's division is preceded by replication of the DNA of its genes with potential for errors in copying. The genes along with some of the regulatory elements and half of the cell machinery and structures are provided each daughter cell. These are all heritable and become available as priors for new biochemical behaviors. Bacterial cells can even share some priors by transferring them as mobile heritable units. This spreads opportunities to disparate taxonomic groups, even Kingdoms. This is the emerging field of horizontal gene transfer imposed on the classical vertical inheritance by trees of descent. Around five percent of the genome within prokaryotes has been acquired through horizontal gene transfer. In eukaryotes the symbiotic fusion of two prokaryotic cells with their genomes and regulatory elements greatly increased the available priors. Living systems build cells with the stockpile of heritable priors as the beginnings of a framework for epigenetic constructions and homeostasis.

The use of existing priors is the subject of Moshe Bar's paper on the proactive brain cited as a reference at the end of the chapter. He suggests that the brain doesn't ask what something is but, "What is it like?" His paper is an excellent introduction to the use of similarities as metaphors in coping with a complex dynamic world of experiences. The paper proposes that the default mode in the brain does a search for sensed features as priors in memories or their similarities as similes, analogies, and metaphors. Speed and accuracy depend on our available priors. We construct a visual model shaping it to match the input, increasing resolution by supplying details in the process. This is also why rereading a difficult book or paper once you know more gives it more meaning and makes it look so different than the first time you read it. Your increased knowledge has allowed you to create a richer and more completely connected set of metaphors or model parts for your interpretation. Since we operate in complex and variable environments, generative models are shaped to incorporate this variability into a probabilistic description of how the input data might have been generated. Perhaps we can consider free energy reduction as the transitional force between metaphor and the new prior. This removes errors and shapes generative models to better fit the purpose. Models represent prior beliefs about salience of environmental causality. Such models can be shaped to predict possible meaning of new sensory data, and thus to infer the appropriate response based on past experience. This functional role over generations of selection serves in neural behavior and also applies to the evolutionary process much discussed as preadaptation or exaptation. Bayesian inference can be seen as the use of existing priors as metaphors for novel input. Those priors that best match the novel input are used to form hypotheses that shape a subsequent response.

Metaphors, models, and gist: how are they related?

We have all experienced another's statement reminding us of a memory of an episode that has some degree of similarity. Recall occurs through the gist of their input triggering our gist of a memory sharing similar features. Does metaphor operate something like gist that provides a higher-level overlap of concept which provides access to useful information for model construction? Since our memories are our personal constructs, their conceptual meanings as well as their perceptual details are individualistic. Two individuals viewing the same episode will construct models of that episode that also reflect their past experiences and values. We can conceptualize this as our behavior being a history of facilitation in neural pathways beginning from similarities in memories. The process would begin by top down selection from the individual's models of their experiences. The model would be refined by bottom up generation of recursive active inference. The neurons that 'fire together, wire together' are the elements of frames that create a script of the story of the new behavior that is interactively shaped. Evaluation of perceptual resemblance between model and task is what the goodness of fit supplies. This is facilitated by selection of the most probable prior to work with. Metaphors at this level reduce to symbols of features as messages for evidence of similarities between neural domains. Selection is a recursive process building representations from a combination of stored features and their refinement by sensory input to create a functional model. This wraps the math, physics, and story into a package. The gist of priors also resembles the memory elements providing a synopsis of the plot of a story.

If we reflect on our discussion in Chapter 6 on the default mode network and its role in autobiographical memory, we get reinforcement for the role of gist being used to create new stories. The hippocampus separates the perceptual details of an episode from the conceptual in memory. The detailed percepts are stored in posterior brain structures and the concepts with personal meaning are in frontal areas. The anterior hippocampus stores the gist of the episode which can access concepts or details for recall or attention. Two individuals viewing the same problem will recall their personal gist relating to their concepts, values, and perceptual details. The conceptual aspects relate to self, themes, values, emotion, and knowledge from our experienced behaviors. The perceptual memories directly support decisions in well-structured previously rehearsed behaviors when they occur in similar context and only require slight adjustment. The conceptual memories are recruited to begin construction of models that will create plans, purpose, and values for new problems, which share thematic similarities with stored memories. Our memories are thus fragmented for storage in dispersed networked areas. Elements involved in different episodes with metaphorical resemblance share facilitated synaptic linkage. Their activation can provide neural elements as beginning priors for construction of models. Memories are not rigidly organized structures but consist of elements for models that an attentional gist can access to meet current demands. Diversity of themes and perceptions and their relational organization by metaphorical similarities provides versatility for their use as priors to assemble functional models to respond to input from a dynamic world.

Life can be looked upon as an individual doing improv on a world stage. Life continually faces the question of what to do next. We have a massive networked set of behavioral programs, acquired through evolution and learning. We use these programs to facilitate decisions and implement behavior in adaptive ways that range from breathing to performing brain surgery. In the general sense, this can all be viewed as behavior of an agent doing something for a purpose and thus it can be treated as cognition. We defined cognition as the use of models to interact with other systems. Structured enzymatic pathways can be models to convert carbohydrates to ATP and neural networks are versatile models to mediate behavior in animals. Starting as infants, we each build complex coherent mental models of the causal structure of our experienced world. *E. coli* comes into its world pre-equipped with behavioral models that worked well in the past. For self-organizing systems to adapt to change and diversity they must continually correct mismatches between their models and the real world in which they live. Our brain uses individual histories to predict how the neural models should respond to novel input in the future. Individuals act to make decisions on sensory predictions come true. Mismatches call for corrections in behavior or models.

Agency is enactive in being dynamically coupled to the environment through sensorimotor activity. Enactivism in cognition refers to behavior being shaped for the purpose of the individual by exploring the environment through sensed challenges and needs. The individual interacting with its environment thus produces its private individual world through its selective choice of perceptions, positions, behaviors, and resources. Elements of neural models from memory most similar to elements of the input are successively assembled from low to high resolution while being adjusted to correct the model to optimal fit. This is the generative model in operation. If it lacks memories to compare with the input, it relates the input to the best analogy that exists until the model can be reworked. This is how we create our story.

Working memory is the active assembly of information to guide decisions in behavior. Humans monitor working memory in their consciousness. The hierarchical and time-sequenced generative assembly of models is similar to the concept of frame, which is the script-like approach described by Minsky and Shank, where models from memory are assembled into a story line to guide behavioral decisions. For a person who has dined out often, past experiences provide memories to guide how you sequence behavior appropriately for a new experience. The physics of any real system is much too complex for organisms to deal with, so only those few properties that best identify and target hidden causality as resources are selected as sensed properties for models. For the activities of each agent there exists a reductionist physical description revealed in physiological and morphological studies of the cell and whole organism. The reductive analysis is only part of the explanation, since there must also be a historical, teleological, adaptive story in order to make the process happen. In a hierarchical agency view of the body there is information used at each level in controlling each agent's behavior. Information about their status is shared between agents controlling integrated behavioral activities in cognitive chains. Networked information flows down, up, and sideways and regulates the behavior of the individual, creating the adaptive self. The cognitive story involves the subjective personal elements—information, knowledge models, purpose, and values—and this story is paralleled by the story of chemistry and physics. The purpose and algorithm exist along with their physical implementation providing complementary aspects of the same reality.

Speak, memory: the multiple roles of memory in shaping behavior.

Some of the information involved in multi-agent systems concerning aspects of the unfolding behavior may be incorporated in memory and stored for future uses. We can combine memories with current events, and like all stories of complex events, the ultimate story that emerges is composed of selected fragments. In humans this process is embedded in networked neural nets that play through working memory and along with our autobiographical memories create stories we replay in our consciousness. What we see, what we feel, and what we know is not only physics and conceptually amenable to a reductionist physical explanation, but it is paralleled by the flow of information related to our knowledge that we have created from our history of experience. This is the story of agency in action that can be told to yourself and others when the **working memory**, involving its images and language, accesses storyline models from our memories to share personal stories. This all derives from the behavior of these models, each structured for an adaptive purpose working symbiotically for the individual. The 100 billion brain cells, each with thousands of synapses, contain vast knowledge in memory of neural nets to construct the wealth of stories of experienced lives. This is the synoptic story of the functioning default mode network we detailed in Chapter 6, which creates our awareness from the hidden processing below our awareness.

"What is it like to be a bat?" would be a story of where to go and what insects to catch. The stories of living organisms can be viewed as improvisational plays and the flow of information in enacting agency. This creates how it feels to behave as distinct from the physical action itself. Remember our adage on the role of physics in life; *Life tells physics how to behave, and the physics tells life how to feel.* The feel referred to represents the purpose and value functions that form the basis of their re-representations we discussed in Chapter 6, as displayed in our autobiographical memory. These are the feelings of life in enactive homeostasis and these functions play roles in the creation and use of generative models of behavior. The story along with its feeling when it is told is embedded in the knowledge-based model that involves the use of information about the structure and causal relations of valued resources. *E. coli* enacts a story beginning with desire for food, perception from sugar receptors, and the flow of information through biochemical pathways in telling its motor how to behave to swim by increments up gradient to a sugar source. In response to satisfying our hunger we could tell the story of how we act to get a sandwich. Acting out stories is the oldest form of storytelling and also helps perfect their meaning through practice and correcting errors.

The story-world was constructed as life evolved and learned, providing the script for adaptive actions. The script was specific to the species and individual. To be conscious is to be an observer of agency in action and what is observed is the feeling of the unfolding of a story. To be conscious in behavior does not require the ability to report your feelings. Remember our batter's feelings about the fast ball and your feelings about driving when unaware of that behavior. The feeling may be the synoptic message of how well the behavior plays out. It may provide the evidence of errors in the behavior and bring this to the attention of the model for correction. A story is the behavior of agents acting for a purpose that has value, playing out in time. The agent is the prime observer of its own story. An adaptive role for the observer is to monitor the progress and success of the story in achieving its intended purpose. This monitoring can be used to make adaptive adjustments in the story so it will be better the next time it needs to be told. This sort of adjustment should apply all the way down to basic acts of agency where cues from the environment that influence its outcome serve as entrance points to the story. On the scale of evolutionary time, such a mechanism can adjust performance of entire biochemical pathways, while on the much smaller time scale of an individual organism, it is equally well-suited to regulate *E.coli*'s lac-operon as it switches between which sugar it metabolizes. The same paradigm applies in a more versatile way to synaptic selection over learning time in neural networks.

Feeling and behavior.

When an agent behaves, it must process a flow of information to perform specific functions. This process creates a story. That story has a unique feel depending on the resulting function. Similar functions or acts, if performed the same way, will feel the same. This feeling informs the agent as to whether or not they are performing the action correctly. It feels the same to do the same act, similar for an act that uses part of the same model. It feels different, however, if it is done incorrectly. Sensing information from another agent has its own individualistic feel. In behavior, meaning results from how it feels as the model judges how well it performed. How it feels plays a role in meaning, value, motivation, and error detection in behavioral knowledge models. This approach as a story unfolds also provides a rationale for the Libet time lag. One cannot consciously respond at the very start of the story. The story must first progress and develop to carry at least a fragment of meaning before it can be considered by our consciousness. Is this the time when the magnitude of its free energy is detected? In chained and hierarchical agency, information relating to either an unfolding story or one that has already played out can be provided to linked agency for use of others.

Life requires and believes in a real world but can only experience it through a reduced selection of information transduced to interact with its behavioral models. Throughout our evolutionary and cultural history, it has been socially important to communicate the experience of 'what it is like' and also to take ownership of 'what it is like'. Folk psychology designates a longstanding cultural tradition of speech, thought, and stories about the mental realm. This discipline has provided us with concepts and vocabulary that eventually transitioned into cognitive studies. Folk psychology and narrative theory have considerable overlap with agency since all three treat behavior. Stories in a basic sense all consist of characters, settings, actions, and events that are linked by time and purpose and often involve conflict and trouble in their resolution. Literary critic Kenneth Burke's *A Grammar of Motives* has an introductory remark, *"What is involved, when we say what people are doing and why they are doing it?"* His dramatistic pentad relates to the elements of a story: Act, Agent, Scene, Agency, and Purpose. The story becomes interesting with Trouble driving the drama. If you substitute 'what life is doing' for people in his summary statement above it clearly overlaps with enactive, embedded, and situated agency. Behavioral models are called upon to implement an action in an environmental setting for a teleological purpose. If there is an error or surprise, the drama of coping and correction appears. Just as is done in storytelling, behaviors can create complex stories when information from one agent interacts with and enhances the story of another. When provided with some form of heritable transmittable memory, both agency and stories can also evolve. We have all experienced how the sharing of humans' autobiographical stories have produced our knowledge that we use to tell our stories. To produce another level of understanding in cognitive process in behavioral models we will explore the connections of narrative and cognitive agency.

Generating a narrative.

Narrative is the story of events driven by the intentional behavior of one or more autonomous agents in a manner that manifests a subjective world of meaningful knowledge and information. This is paralleled by the events played out by the actors in the physical world. Let us regard stories as causal behavior playing out in time, which have values and purpose. Narrative production is the behavior of the agent or its model playing its adaptive role in enactive communication to get that behavior optimized. However, if the agent is participating in a multi-agent system, it is also providing information about its behavior for the use of other agents. There are two stories in an act of agency: the one we tell ourselves and the one we provide as information to influence the behavior of other agents. Much behavior is scripted but behaviors are open to improvisation when meeting with challenge or opportunity.

Reflect on the discussion of effective field theory from Ch. 4 as we consider this treatment of life as a story. The physics and chemistry of living systems fit with reductionist explanations, but agency provided a story of interactions based on history. Science grew to expect and respect the approach to scientific explanations based on reductionism and predictability. H Porter Abbott is a narratologist that not only treats narrative theory in the arts but also the role it plays in people's lives and the communication of information. I first encountered his work when I was reading a paper where he described evolution as complex in so many dimensions and process that as a story it was unnarratable knowledge. This was as opposed to the creation story of the Bible that people easily related to. He pointed out that in complex systems where untold numbers of micro events produce macro events, with contingencies at many levels and time scales of billions of years, reductive approaches to explanations are neither feasible nor rewarding. These systems are best explained by extracted reductionist stories at micro and macro levels so that a holistic narrative theme can be developed to provide perspective on the system. We first need the Big Bang to cosmology stories to get our planet. An example in evolution is the horse story from the dog-sized forest dweller with four toes to the large modern horse of the open grasslands with one toe as the hoof. Mendel making crosses of pea varieties to fruit fly genes to DNA makes RNA makes protein is another part of evolution that is a story. There are millions of stories and how they are selected and sequenced determines their explanatory value. Bohr informed us that physicists tell us stories: *"It is wrong to think that the task of physics is to find out how nature is. Physics concerns what we can say about Nature."* Levels of complexity with emergent properties are difficult to explain since they are not revealed by a reductive causal analysis and properties emerge with the construction of the higher level.

Living systems (individuals) are constantly changing in both their internal agency and the relation of the individual to its dynamic environment. Internal changes create different priorities and the external changes create opportunities or challenges. Internal changes that lead to an exploratory phase in *E. coli* could come from reduced phosphate concentration at the flagellar motor switch that in turn puts it in its run mode. Movement during the run through the environment will generate signals at its receptors, which may in turn bias its trajectory based on alterations of the frequency of runs and flops that move the cell to favorable conditions. Movement up a sugar gradient will result in the uptake of nutrients and provide substrate for the agency of metabolic pathways. The micro-actions of enzymatic pathways will break down the sugars, reducing them to CO_2 and H_2O and supplying ATP as energy currency to keep the story going. Scientists in laboratories have studied many of the elements of what *E. coli* is doing and have detailed stories available to those who read and understand their papers about genes, molecular biology, chemistry, and physics of the agency of micro-events supporting these actions. Science uses reductive approaches to explain or describe the causal nature of observable events in the behavior of the cell. These have an adaptive causal cognitive basis that we can link together in a storyline for a causal purposeful scenario.

Our stories in myth, literature, and film all share a familiar structure throughout human culture. A hero is called to an adventure, has some success, then is faced with frustration and danger, but when all seems lost finally prevails and succeeds in his or her quest. This is a basic story of life involving the evolution and behavior of environmental interactions in agency. This is also cognition in the sense of using models and energy in work cycles to interact in an adaptive way with another system, which presents challenges. This basic theme is revealed in our bacterial chemotaxis story where *E. coli* is faced with an energy crisis.

A young *E. coli* much involved in its routine homeostasis feels its supply of ATP declining and anticipates it must seek a source of some energy-rich substrate as a metabolite. It begins its cyclic flagellar beat, ventures forth and boldly begins to explore. It is on a quest of discovery for sugars and amino acids and has innate confidence that if it persists it will succeed. It searches persistently and randomly but then encounters dangerous and toxic conditions and must correct its course to avoid these hazards. As its journey leads to friendlier conditions other dangers loom and it barely escapes the feeding vortex of a hungry ciliate. Finally, after minutes of its quest, it picks up traces of the scent of glucose and launches a search for the source. It swims up the gradient, reaches the mother lode and replenishes its ATP. It then puts this to work in growth, which stretches its skin and culminates in cell division and becoming twins, which then swim off together before each chooses its own path and begin their separate journeys.

It is a long way from bacterial chemotaxis to Homer's *Odyssey* and it took three billion years to get there. What we recognize as stories are always behavior in action for a purpose, which is the product of an agent's programs trying to make things better when presented with a surprise or challenge. Life's story plot is decisions on what to do next and our *E. coli* did a successful journey when faced with challenges. It did all the right things for itself, cope with its environment and provide for the future. *E. coli* had different values and goals than Odysseus but with poetic license its voyage could expand to a dramatic story or a captivating film.

More is different.

We have heavily employed Sean Carroll's poetic naturalism in our story telling of cognition. It helps to synthesize and understand the functions that accompany the physics of behavior. Again, Niels Bohr provides help:

> "Quantum provides us with a striking illustration of the fact that though we can fully understand a connection ... we can only speak of it in images and parables. We must be clear that when it comes to atoms, language can be used only as in poetry. The poet, too, is not nearly so concerned with describing facts as with creating images and establishing mental connections."
> Physics and Beyond: Encounters and Conversations (1971)

Sean Carroll cautions us to keep our connections between levels correct when we use poetic naturalism. We have introduced functions at cognitive levels as having new emergent properties that were not present in the physics before it became part of the living behaving system. The poetic naturalism stories of classical physics and psychology are accounts that provide understanding of complex systems for theoretical treatments at higher levels of interest [without necessarily concerning ourselves with what is going on a lower levels]. It is the emergence of new properties that makes it necessary to use poetic naturalism in order to investigate and explain these complex systems. As materialists, we recognize the basic microstructure of the world with fermions and bosons and relate stories about quarks and gluons to move up a level. As we move to higher levels, the stories relate to properties and functions that we assign first to the atomic level and then to the molecular level. We build our classical world of physics at this level and follow its aggregations into the material composition of the universe. When I became interested in complexity, I read a helpful paper by Philip Anderson. In a 1972 paper, he pointed out that, *"More is Different"* and that complex systems often exhibited emergent properties that eluded reductive explanations. For analysis, this new system required new fundamental principles at those higher levels and new disciplines were required in order to understand it. These are the effective theories for emergent levels.

In the book *"The Algorithmic Origins of Life"* by Sara Imari Walker and Paul C. W. Davies, the authors propose that the role of information in the origin of life was to direct process in structured biochemistry.

> *"Here, we propose that the emergence of life may correspond to a physical transition associated with a shift in the causal structure, where information gains direct and context-dependent causal efficacy over the matter in which it is instantiated. . . . This leads to a very different, context-dependent, causal narrative— with causal influences running both up and down the hierarchy of structure of biological systems."*

My major difference with the authors is that they place information control coming before the emergence of knowledge and meaning, while I incorporate them in a recursive evolutionary interaction. I also applaud their put down of reductionist biology and its assumption that ultimately all life is nothing but chemistry, since I conceive biochemical pathways as evolved behavior all the way down. *"How does a molecule become a message?"* posed by Pattee in 1969 as the title of a publication, suggests that along with the reality of its atoms, a molecule now has acquired a new functional role in a behavioral system.

Emergent properties: realities and concepts.

The question in philosophy becomes, do the cognitive functions also constitute real, or just conceptual explanations? Are the functions causal or are they simply allegories? These questions steer us into metaphysics, which was always esoteric but now reaches enigma proportions. Before the quantum world, the Big Bang, black holes, multiple universes, and virtual reality, we were comfortable with the reality of classical physics. Now, we don't really understand how even that comes about. If the universe is nothing but a huge wave function, I am not sure of what its properties are. I can't describe what energy is like, but I am told it gives rise to forces, particles and their interactions and this is responsible for the world of physics. I also fail at a rigorous description of the metaphysics of consciousness. It may be that like energy, space, time, force, and fields used in physical process, consciousness can assume many forms and it is a process rather than a thing. However, I would make the claim that consciousness is what gives rise to information, meaning, and their interactions and these interactions are what has produced the complex world of life on Earth. Most would treat my statement with skepticism. However, Erwin Schrodinger (1887-1961) used the existence of autonomous behavior as a way to recognize the difference between life and physics. Behavior is based on the flow of information that directs interventions on physical systems and keeps feeding on negative entropy. Thus, life is the production of a physical knowledge model that interacts with another system/other systems for its own support and maintenance. This is what ultimately provided me with my reality for the observer role of consciousness.

Now we must contend with both the science and philosophy of emergent properties and whether these properties are real or conceptual. The vast majority of the existing controversy over cognitive properties is centered on humans and similar animals, as if these were emergent at some level of neuronal activity. We can understand principles of behavior by treating similar functions at lower levels of evolution and development where processes are observable and tractable. This is the usual process of science for analysis of complex systems. I would make the case that if the emergence of life, and the behavior that it is based on, are real, then its new emergent control functions are real.

Control theory tells us that observers must acquire information about another system in order to interact with it. That information must have meaning in order to participate in shaping an adaptive behavioral model of the environmental system that permits predictive interaction with it. The observer and its representation of the environment are parts of the world along with the system itself that is modeled. Together all three constitute the observer's world. Environmental energy from a resource becomes information when transduced by the observer into a biotic symbol with meaning that communicates a message. How does Howard Pattee's molecule acquire the message it carries? In a quote from Sara Walker and Paul Davies's book, "*information gains direct and context-dependent causal efficacy over the matter in which it is instantiated.*" This information did not exist until evolution provided its origin and role in biochemical processes. In his book, *What is Life?*, Schrodinger tells us that we can tell when life is present because the system in question keeps behaving; that truism has been the on-going story since the first cell, LUCA; 3.8 billion years. If an observer is required for the transformation of energy into a biotic symbol with meaning, the observer would seem to be a real part of the system. Our preceding chapters have presented the case that shaping, correcting, and controlling the behavioral model requires the functions ascribed to consciousness. To provide meaning for the information, consciousness also requires other cognitive functions in both evolutionary and learning cycles.

What consciousness really is becomes relevant when it functions in our free will decisions. Free will plays a huge role in individuals and societies. We discussed this in Chapter 6 in relation to the default mode on pages 13 and 14. The consciousness everyone talks about in free will decisions assumes their ongoing display is the decision making in process. We present consciousness as our display being a sample of the background processing called to attention for more background processing to approve or disapprove decisions. The attentional input can supply evidence to guide actions and perhaps make selections for what to store as memories. This is a re-representation or meta-consciousness as a virtual display of the latent representations of the information processing in consciousness. The display can combine memories with extracts from sensory input.

Libet's half second delay brought up the question of the existence of free will and Libet decided that what humans had was "free won't," since action could be inhibited. However, the free won't also resides in the background decision making. This background processing is the functional role of the unaware consciousness working in generative models that will then lead to behavioral interactions for more input calling for the cycle of recursive processing. Our free will is experienced in our decisions in behavior. These decisions are shaped by our personal history choices while processing information as evidence in Bayesian inference. Your selection of priors as memories will be used as values and purposes established by previous behaviors. Only by questioning beliefs, values and purpose can you improve or change behaviors, thus the reality and functioning of the cognitive functions of generative models is relevant to living systems.

These biological emergent properties that we use in cognitive functions and behavioral studies are treated as being subjective since their expressions are unobservable and individually variable. Do they provide functional roles that the underlying physics cannot do? The underlying physical structures in the living system and the roles they perform are not predictable from the laws of physics. The properties are referred to as subjective rather than objective. Bertrand Russell reflected on this subjectivity in the 1950s in his essay *"Mind and Matter"*: *"We know nothing about the intrinsic quality of physical events, except when these are mental events that we directly experience."* From conscious experience, we learn about the properties of the physical world through the action of our assembled knowledge models. We assign physical properties based on their predictability and sharing of observations and measurements renders them objective. Physicists can't predict the cell using just the rules of physics and predictability decreases with increasing behavioral complexity

Physics works autonomously, following long-established rules. We introduced behavior as an adaptation to intervene in ongoing physics in order to change the outcome. Life does this in a variable world of contingency and choice. Life's interventions are probabilistic with causality, context, purpose, and decisions. The more complex a living system is, the less predictable it becomes. When considered this way, life itself may be viewed as subjective down to its very origins; but life is also the only reality that we have. Through science we have created detailed understanding of physical and biological worlds, both made from the same matter. To understand living systems we added purpose, meaning, and control by making life's path reliant on adaptive behavior. My experience has created the only world I know as real, though it is quite messy and unpredictable. I continually strive to learn more about it and correct errors to enhance my understanding. The appearance of life on earth introduced a broad spectrum of outcomes and interactions to adapt to diverse experienced environments. Our chapters have examined similarities in function and process in the spectrum of behaviors from life's origins through its diversity as each line developed its own story.

Building and directing one's life story: a summary.

Life weaves complex stories throughout living systems, which have undergone learning and adaptation during eons of selection on behavioral traits. These behaviors participate in so many interactions within and between systems that it boggles the mind to understand how life copes. To have a future an organism must be good at correctly deciding what to do next in a challenging, dynamic environment. This is what we have tried to explain with the hierarchical assembly of generative models that employ Bayesian processes.

Since we live our lives as participants in stories, we should concern ourselves with how we write the script. The story will be written and acted out by the coordinated interaction of behaviors generated by our hierarchical models. This process determines who we are. We write and modify the script of our lives with the aid of the consciousness as we live, select, and interweave autobiographical memories. Our consciousness should be involved with the direction of our story and improving the script. This improvement can be conceived of as creating the documentary film of one's life. Consciousness is the director—in charge of the production team— assuring a quality product. This role was presented in the previous chapter as the interaction of consciousness with the default mode network DMN. Consciousness must be well informed and wise in decision-making since it works at 40 bits per second compared to hundreds of millions by the DMN. If you have ever wondered why the trailers for credits of participants in making a movie are so long, it is because a very complex process is involved in what we see as coherent behavior on the screen.

What we see when we screen our personal documentaries are clips that our consciousness, in the role of the film's director, must edit. These are only selected fragments of a story we are creating and telling to ourselves. The production crew (as communicating neural models in the DMN) direct the focus of attention and play the clip for the director. The quality of the film depends on selecting high priority fragments as clips for decisions on editing. Stories have gist and gist provides a sense of knowing the plot elements, where you are, and where you can go. As we watch our film, attention evokes the gist which provides clues to errors and opportunities for processing to the DMN to begin the model for the next behavior to improve the story line. It requires a lot of experience to be good at choosing what to do next. Solving the problem of getting the right stuff is facilitated by cross-referencing memories with their potential relatedness. As a filmmaker, this process is accomplished by logging film clips. We do this logging by laying down memories as context and details with gist to relate them.

Prepare to think broadly. This is how we get priors for our Bayesian inference. They begin with attention of observed input, which seeks to find the simplest and most likely explanation as it builds a behavior model. This model building selects from our knowledge and is optimized by recursive observations and actions. Audience approval and awards will relate to the meaning we build into our story, but in evolution this success determines survival. If we weave a story lacking in virtues and social values, though we may get benefits, it will contribute to societal problems.

The moral of the story.

Remember, my approach to cognition takes me back to its basic origins in development and evolution. The story concept applies all the way down, beginning at the level of biochemical simplicity where we can analyze and understand biochemical evolution and function. We see that stories at this level permit understanding of chemotaxis and regulation of the biochemistry to run a cell through its life cycle. We can see the functions but how they feel remains an enigma. However, I began my cognitive studies to understand issues relating to the meaning of evidence and decisions since I was interested in the role of morality and justice in life. I thank Wikipedia for the concise statement on a complex topic. *"A **moral** (from Latin moralis) is a message that is conveyed or a lesson to be learned from a story or event. The moral may be left to the hearer, reader, or viewer to determine for themselves, or may be explicitly encapsulated in a maxim. A moral is a lesson in a story or in real life."* My view is that life has learned lessons well in creating its story, that it is very moral and evolved by using experiences from history to predict the future. My next chapter treats living a virtuous life viewed from the standpoint of life's evolution.

Further Reading and References.

Anderson, P. (1972). More is Different. *Science.* 177: 393-396.

Bar, M. (2007). The proactive brain: using analogies and associations to generate predictions. *TRENDS in Cognitive Sciences* 11: 280-289.

Lakoff, G. (1992) *The Contemporary Theory of Metaphor.* in Ortony, A. (ed.) *Metaphor and Thought* (2nd edition), Cambridge: Cambridge University Press.

Gabora, L., E. O. Scott and S. Kauffman (2013). A Quantum Model of Exaptation: Incorporating Potentiality into Evolutionary Theory. *Progress in Biophysics & Molecular Biology,* 113: 108-116. https://doi.org/10.1016/j.pbiomolbio.2013.03.012

Pattee, H. H. (1969). "How does a molecule become a message?" *Developmental Biology Supplement* 3: 1-16.

Russell, B. (1956). "Mind and Matter", In: *Portraits from Memory.* London: George Allen & Unwin Ltd.

Walker SI, Davies PCW. (2013). *The algorithmic origins of life. Journal of the Royal Society Interface* 10: 20120869. https://doi.org/10.1098/rsif.2012.0869

CHAPTER EIGHT

Virtues, Evolution and Life

"What we need and what we want is to moralize politics, not to politicize morals." — Karl Popper from The Open Society and Its Enemies

Most people treat morals as emergent in human societies, but my question focuses on how morals could have evolved and developed throughout evolution, eons before humans, and how their development relates to adaptation. Morality applies to our topic of behavior and its consequences. We have stressed the importance of behavior leading to a better future and I had become concerned for the Earth's future. The lack of action on our environmental problems led to my concerns on intergenerational justice and our responsibilities to the future. Thus, I became curious about what justice really is and how justice relates to cognition. Since justice is a concept that relates to individuals, society, philosophy, religion, law, government, and laps over into animal and environmental rights, there has been a plethora of academic treatments on its definition/description. I found that justice was first described as one of the virtues by Plato and was probably discussed by Socrates, but he didn't write down his thoughts, leaving that task to others, notably Plato. Virtue is behavior showing high moral standards. There are four cardinal virtues. The cardinal virtues are so called because they are regarded as the basic virtues required for a virtuous life and are involved in behavior.

Religions prescribe values to guide adherents in their personal behavior and the Bible adopted Plato's virtues. I found this on the CERC site.... Catholic Education Resource Center, "*The four cardinal virtues – justice, wisdom (prudence), courage (fortitude), and moderation (self-control, temperance) – come not just from Plato or Greek philosophy. You will find them in Scripture. They are knowable by human nature, which God designed, not Plato. Plato first formulated them, but he did for virtue only what Newton did for motion: he discovered and tabulated its own inherent foundational laws.*"

The Cardinal Virtues have a deep evolutionary history.

We now believe Newton's laws of motion should be understood from the perspective of the Big Bang, quantum events, and the emergence of our classical physical world from quantum mechanics rather than a deity. Ethics and morality derive from many sources and many cultures. Western culture grew from the Greek philosophers. Early Christian scholars incorporated aspects of their thoughts in their theological doctrines. There remains a strong religious association with morals and ethics in society. For our classical ethical values, their definition, origin, existence, and function have an immense and complex history in philosophy, religion, psychology, law, society and the role of ethics in the lives of individuals. These cardinal virtues are properties attributed to living entities and their behaviors and our question concerns their origins and roles in the evolution of life. The treatment of ethics and morality have the same association with high level neural activity as consciousness, which we have treated as to origins and functions.

The four virtues are described as cardinal virtues since they are regarded as the basic virtues required for a virtuous life. Other morals and ethics should be derivable from them. I will simplify by regarding virtue in behavior and examine life's history and properties to see if we can relate virtue to biological processes. This point of view is given philosophical justification by Wittgenstein, who said that morality and ethics were not things that you talked about but things that you did in his work, *Tractatus Logico-Philosophicus*,

What if we translate the virtues into properties of functions to see if they could have served important roles in evolution and cognition? *Justice* can be simplified and considered as fairness. As a cognitive biologist I might assume justice is involved in predicting the most adaptive decision and action by assessing all available information. This process seems somewhat universal.

In the big picture, this definition could cover fairness, equity, rights, and even law in a democracy. I am more comfortable reflecting on *wisdom* and describe it as the knowledge models for appropriate behavior in response to the available evidence. Even celebrity British spy action movie star, James Bond, may agree that *courage* represents the decision and motivation to act when presented with a challenge. *Moderation* seems to fit the concept of homeostasis, which is maintaining equilibrium and stability when challenged by options. It has long been accepted that justice was the most important virtue. In my analysis of the others, I fit them under my umbrella treatment of justice. A fundamental truth in biological systems is that properties of living systems are interrelated. This is true for properties of biochemical pathways run by genetics and epigenetics and networks of models in neural pathways. Interrelatedness should carry a deep message to societal and cultural expressions of these virtues. Justice without wisdom, moderation, and courage is a problem for a democratic society.

Since values in living systems are interconnected, justice facilitates symbiosis or evolved social behavior. Justice can be regarded as the value derived from the most equitable long-term interactions between the agents of a multi-agent system. The success of the system derives from justice for all in the evolutionary court. This covers life from metabolic pathways and epigenetic interactions in gene regulation to relations between international societies. Wisdom is Bayesian optimized inferential behavioral models of living systems. This idea is detailed in the writings of Karl Friston on generative models for behavioral decisions. The model optimizes knowledge for prediction of best behavior to solve a problem.

Friston employs free energy as a metric to measure the discrepancy (error) between a model's predictions and the actual outcome. Knowledge is improved by practice, openness to evidence, and error correction. Courage is an adaptive response to perceived threats or challenges. Courage is also required to perform the task of resolving all details of problems that arise. Moderation resembles homeostasis by detecting where and when to do something to keep the system in a balanced equitable state, so it is ready to respond if challenged. Solving problems as they arise prepares you to face the future.

Justice is achieved through information exchange between agents to regulate and optimize their interactive behaviors in adaptive equitable ways for all. Think of the cell as a chemical factory that is running, repairing, and producing products as it constantly balances dynamically changing demands for each of those products. Wisdom results from learning via Bayesian optimization by models that predict adaptive interventions into ongoing physical events and communicate between models in management for the future. Courage is the initiation of adaptive action to cope with the challenge of a complex dynamic environment. Moderation is the use of these virtuous acts to use energy and resources effectively. Moderation balances opportunities and potential harm in order to choose appropriate behaviors that best maintain the life of the individual.

If virtues have been functionally present and acting throughout evolution, how and why do they emerge as discrete concepts as humans acquired language and culture? Wouldn't our lives benefit if we just keep doing them rather than describing, ignoring, violating, or redefining them? Their emergence in human cognition as descriptions of functional roles throughout evolution perhaps relates to the importance of social organization in multi-agent systems. Perhaps virtues as an attribute of living systems would only be a philosophical, cultural, semantic discussion if these virtues were not so important in our society. Are our defined virtues real things playing a functional role or are they simply rationalizations of behaviors and not motivators of virtuous behavior? In this chapter virtues are described as playing important roles for billions of years in evolution and question what we can learn from their history for a more virtuous society.

In its origin and evolution, life created properties and functions by conducting experiments in physics. Life used measurements to construct, correct and refine models in order to understand and predict the results of targeted interactions that serve adaptive purposes. Life's wisdom, knowledge, information, purpose, and values are employed to create and understand the causality that uses physics for adaptive interactions. We create explanations of why and how the physical entities mediate this behavior. These virtues function in the cognitive processes that parallel the actions of physical entities. Justice assures that *wisdom* is used with *courage* to achieve the *moderation* of homeostasis, which in turn allows an organism to keep behaving justly. Behavior with *justice* is the primary goal of living systems. Life can only exist when it behaves in a manner that maintains itself within the boundaries compatible with its capabilities, which is the concept of homeostasis. Virtues are functionally interrelated and ultimately relate to predictability in behavior of multi-agent systems.

Before the origin of life, matter and energy acted with the inevitability predicted by physics' rules. Life created contingent actions so that different choices would have different results and consequences. With choice, there could be good and bad behavior. This is when the virtues appeared to play their role in optimizing decisions. Life with purpose, goals, values, adaptation, decisions, and rewards that operated for nearly 4 billion years provides relevance for modern day concepts of virtues. However, life doesn't talk about it, write it down, put it to equations nor provide definitions. That is what we humans have done with language and culture to create our manifest as well as our scientific and philosophic views of the world. As cognitive socially interactive humans, in order to communicate we must share experience to agree on the structure and function in our common environment. With physical objects we can share verbal symbols and agree on the assignment of their 'properties'. Our subjectively experienced qualities are more ineffable. They produce ambiguities in shared definitions. This difference between physical reality and subjective experience becomes particularly problematic when individuals or groups are isolated from one another. The definition, usage, and translation of these mental constructs can diverge, making mutual understanding more difficult or impossible.

Stories: physics versus biology.

Counterfactuals and behavior all create stories and stories have motives, ethics, right and wrong, and emphasize different interests. I believe that physics' story is told through unambiguous mathematics, but that life's story is one with historical contingencies. Our personal autobiographical story structures our participation in the grand story of life. This is of course where morals and ethics get messy. In evolution I described virtues working with knowledge, information, purpose, and error correction to create models for behavior in living systems. For humans, 37 trillion cells of 256 different cell types, each with their individual histories and idiosyncrasies, structure our bodies. These components have worked out their moral and ethical issues rather well and function as sets of societies of individual cells working cooperatively to carry out challenging assignments while also responding in coordinated functional ways to requests and commands from higher authorities. This process creates a rational whole with information flowing up, down, and sideways without even knowing who is in charge or if anyone is. The whole operation is just an assemblage of agents working moment by moment to make predictions of what to do next based on the available evidence. We are the result of nearly four billion years of these virtues interacting with challenges and opportunities offered by the environment. As civilization emerged, our cultural discovery and adoption of these virtues challenged us to continue to employ them in adaptive ways.

Justice: the root virtue.

To create a functional society, we should subsume the other virtues under *justice*, which stands out as the fundamental virtue and could facilitate derivation of all other virtues. The American moral and political philosopher John Rawls (1921-2002) speaks to the creation of a *just* society through the favored trick of philosophers, the thought experiment. If the knowledge and values of all members of society were used in planning its governing rules without having knowledge of any individual's identity, rules should be both fair and functional for all and provide for the future. This approach uses a veil of ignorance in order to remove self-interest, greed, and bias from the use of knowledge and values. These selfish motives would be separated and recused from participation in deciding how to make justice and equity universal. The Roman emperor and Stoic philosopher Marcus Aurelius (121-181) wrote about the different levels of goodness. The first, he said, was being good in order to put someone in your debt. The second level was to be good in order to feel good about yourself. The third, or highest, level was to be good like grapes on the vine. They don't grow in order to feed birds or humans; they just grow because that's what they do. If justice were built in by evolution, why don't we just do it for the sake of fulfilling our evolutionary potential for a virtuous world? As a cognitive evolutionary biologist, I would inform Marcus they fulfilled all the virtues as they grew to feel good and make grapes to feed others as part of their lives. The grapes behave as part of an interactive community. Why don't we behave to provide the best and most equitable future for the interactive and interdependent world of life?

Our studies have revealed many lessons of success from what could be taught as morality stories from evolutionary history. We earlier pointed out that 'altruistic genes' rather than 'selfish' could best explain their longevity. Genes work as cooperative team members in their adaptive epigenetic expression in traits. Some have persisted for over 3.5 billion years from their appearance in LUCA, the ancestor of all cells. Adaptations that lead to enhanced fitness require the cooperative actions of many genes under the control of epigenetic behaviors controlling their expression. Early in evolution, bacteria discovered that they could enhance the outcome of behaviors among members of a group by communicating with molecular messages released into the environment. Those messages coordinated the expression of specific genes and behaviors which facilitated group processes. This quorum sensing occurs both within and between species. It facilitates social processes such as: group feeding, producing biofilms, multispecies aggregation in symbiotic mats, horizontal transfer of genes, defense, production of bioluminescence, and reproductive processes. There are many detailed stories for these group processes, but each one is based on the basic principles of multi-agent systems adaptively communicating for the benefit of the group.

One fascinating story of how working together enhances the biotic world is the emergence of the eukaryotic cell. Life above the bacterial level is based on a cell with a membrane-enclosed genome of multiple chromosomes and mitochondria in its cytoplasm. This cell came from two species of microbes—a bacterium and an archaeon—from different Kingdoms with mutually compatible metabolisms fusing into one cell. With the added genetic diversity and later symbiosis with a photosynthetic bacterium, this eukaryote ancestor rapidly diversified into more than three dozen protist groups, which included those that would give rise to plants, animals, and fungi. These three major multicellular groups with complex body plans have adaptively specialized cells, all with the same genome. Genetic regulation has specialized cell use for different functions, thus cooperating for the whole individual above their own interests. This strategy required the cells to communicate with each other and between different specialized groups. Communication allowed for coordination. The first process included developmental unfolding of the cells, then their internal maintenance and interactions with an ever-changing environment.

Earlier in this chapter I emphasized the intelligent adaptive governance agreed on by the 37 trillion cells of our bodies, and I would add with remorse that the 330 million humans in this country seem totally inept at governance in 2020. A cancer cell is an example of the violation of this communication to maintain cooperative activity when self-interest drives its behavior to outcompete and acquire resources for its growth. Its proliferation comes at high cost to its host. Evolution provided a gene to help avoid this difficulty by getting active when the genetic machinery of a cell begins to act inappropriately. One example is the problems that begin when the *TP53* gene produces the tumor protein (p53) that stops gene replication and begins genetic repair of the cell. If these steps prove ineffective, the p53 protein tells the cell to commit suicide. We have one copy of the gene and about two thirds of our cancers have had that gene inactivated. Elephants are long lived with bodies that have many large populations of replicating cells. To avoid the inactivation of this cancer-preventing gene, elephants have replicated it 19 times in their genome and thus experience very few cancers. Some other large long lived mammals have also replicated the gene. Human civilization has provided for a longer life span as well as the creation of many new chemical compounds that disrupt the normal functioning of our genes and can lead to cancers; all of these environmental changes occurred faster than evolution could respond with defense against cancer similar to that of elephants.

Life cycles that protect offspring with parental care are obvious cases of moral behavior that reflect a concern for the future. The Edge Foundation is a science based organization that poses an annual question to a broad selection of expertise. Alison Gopnik astounded me with a response to an *Edge* question that included the statement that the only two species of mammal with postmenopausal females were humans and killer whales. Since then three more species of whales have been added to the limited list of post reproductive female mammals. She expounds on this in her book, *The Gardener and the Carpenter*, where she makes the case for the role of the wise caring grandmother early in human evolution being of critical importance to group rearing of children and the sharing of her wisdom. She also points out recent fads in parenting designed to shape the lives of children that would be best replaced with the philosophy of the gardener that commits to helping things grow best through support rather than trying to mold them to a design.

The morality of evolution.

The stories above are a few selected examples of what could be lengthened into a book. The stories are examples to demonstrate that instead of evolution being viewed as *red in tooth and claw*, the important role of cooperation in multi-agent systems creates a better future for all life. What can evolution tell us that will help society function more rationally and prepare us for a better future? I would suggest the answer is to access the morality inherent in open communication that leads to just decisions based on all the evidence. The networking of biochemistry within a single cell and the remarkable coordination of actions of all the disparate cells of our body into a functional individual, all speak to the importance of agents democratically sharing their view of how the world should work. This is rather like Karl Popper's Open Society as a marketplace of ideas to work out agreements on the best solutions. Communication is all based on information exchange and Bayesian adjustments of knowledge. The importance of shared communication in the shared system of interactive agents is critical to the functioning of that system. Each agent has its own view and priorities, but to work effectively as a group each must share information and compromise its goals in order for the system to function and prepare for the future. If adaptive coordination of beliefs is not achieved, the system is destined for failure and becomes a dead twig on the tree of life.

Executive function and just behavior.

In an article titled, *"Neural correlates of maintaining one's political beliefs in the face of counterevidence"*, the American philosopher and neuroscientist Sam Harris (b. 1967) and collaborators begin their introduction with this thoughtful statement:

> *"Few things are as fundamental to human progress as our ability to arrive at a shared understanding of the world. The advancement of science depends on this, as does the accumulation of cultural knowledge in general. Every collaboration, whether in the solitude of a marriage or in a formal alliance between nations, requires that the beliefs of those involved remain open to mutual influence through conversation. Data on any topic—from climate science to epidemiology—must first be successfully communicated and believed before it [they] can inform personal behavior or public policy. Viewed in this light, the inability to change another person's mind through evidence and argument, or to have one's own mind changed in turn, stands out as a problem of great societal importance. Both human knowledge and human cooperation depend upon such feats of cognitive and emotional flexibility."*

They found that challenges to political beliefs increased activity in the default mode network, which then activated the executive function to deal with the posed problem. Their study only involved politically liberal people and stressed that the executive function of consciousness gives access to the broad spectrum of knowledge that resides in the unconscious self, which in turn allows the consciousness to accept evidence, modify beliefs, and approach agreement. If the executive function is not accessed, the conscious self will maintain the belief rather than attempt any social adjustment.

The ability to see opportunities in emerging perceptions is the charge of the executive function, which operates in conjunction with our conscious working memory. This system is connected to our background of knowledge that helps build behavioral approaches to problem solving. The executive function is operational in those individuals who reject quick automatic responses and instead engage their frontal lobes to think about creative behavioral solutions that yield the most adaptive outcomes. The following is a list of abilities/qualities that emerge and accompany exercising one's executive function.

1. control of attention,
2. control of inhibition,
3. emotional control,
4. override automatic response,
5. deal with novel situations,
6. counteract biases,
7. self-control
8. planning,
9. organization,
10. flexible thinking,
11. learn and correct errors,
12. create visions of the future,
13. solve problems,
14. motivation to action,
15. accept responsibility for actions,
16. don't blame others,
17. compromise.

This long and impressive list comprises the qualities that can lead to a virtuous life. They can make you creative, functional in a group effort, responsible in an emergency, happy and productive in career and marriage, and a responsible member of a democratic society. We hope to find these qualities in our leaders and decision makers in a democracy. These qualities are associated with activity of our prefrontal lobes interacting with other brain areas all mediated through activities involved with the default mode of the brain.

The prefrontal cortex creates and strengthens network connectivity with other areas of the brain that include the default mode network (DMN) where unconscious processing is enhanced when focused conscious activity is attenuated. The DMN was treated in depth in Chapter 6 on deep learning and Chapter 7 on stories. The DMN is proposed to be active when individuals are engaged in internally focused tasks including autobiographical memory retrieval, envisioning the future, and conceiving the perspectives of others. It has been suggested that the DMN works with the executive function and working memory to share information between brain regions with related knowledge. Relating different themes can enhance finding new perspectives for difficult problems. It is heavily involved in social and empathic interactions with others. When does attention to input get routed to higher-level contemplative treatment and when is it relegated to quick emotional decision? Recent comparative studies of the brains of people with different political stances help to provide a partial answer to this question.

How do brains of liberals and conservatives differ?

A paper by Ryota Kanai *et al.*, *Political Orientations Are Correlated with Brain Structure in Young Adults*, provides some evidence on the question. The study made measurements of grey matter volume in individuals that placed themselves on a liberal to conservative scale. The study shows highly significant differences in anterior cingulate cortex (ACC) and amygdala structure. Liberals had higher grey matter volumes in the anterior cingulate cortex (ACC) while conservatives had higher values in the amygdala. The ACC is identified as the gateway to the executive function. A detailed discussion of this is in a paper by Vinod Menon and Lucina Q. Uddin, *Saliency, switching, attention, and control*. The ACC is a relay in high-level processing of error monitoring and action planning. It also, along with the anterior insula (AI), plays a role in empathy-related responses and uses contextual cues relating to fairness and group identity in decision-making. Input into the ACC and AI begins the executive role in detecting the salience of change and motivates subsequent shifts in cognitive behavior. Sensory and emotional inputs act through this hub, interacting with background information. The background relates ongoing input to saliency concerning contents of the DMN. The AI and ACC act as an attentional hub for cognitive processing in linking the DMN to frontal cortical areas in executive functioning.

The amygdala is involved in emotion processing and is especially reactive to fear-inducing stimuli. A region of the amygdala has been associated with fear conditioning where learned associations are stored in neural circuits. When input associated with a learned fear stimulus occurs, it activates the expression of defensive aversive responses. The input is associated with a learned fearful bias and there is no need for conscious feelings of fear to intervene. This enables conservatives in building group identity and share a history of fears with a vocabulary of words and concepts interacting quickly with ideological agendas.

Human judgment and decision-making are heavily influenced by a broad array of cognitive, perceptual, and motivational biases, of which we are unaware since they work unconsciously. However, we easily detect these in action in others. People make the assumption that they see the world as it is, while others distort it through action of their individual biases. These different views of the world lead to problems for interpersonal and intergroup communication and arbitration. Differing beliefs regarding important problems, and how they are unconsciously influenced, present a huge challenge to reaching consensus on rational solutions to those problems.

On what factors and processes do common beliefs depend?

A belief is a knowledge model that enables interaction with the environment. It also works to reshape other beliefs in your mind and the minds of others. Since communication is how we create and structure society, we must consider what is involved in the transition from communication to understanding that can lead to action. Knowledge resides in models that are able to use information for adaptive behavior. Knowledge is built into the behavioral models in the minds of individuals and it is information relating to that knowledge that we can share. Communication is social interaction based on selection of packets of information from your knowledge models of your beliefs and externalizing them through some appropriate behavior as a message you wish to convey. Reception of the message depends on how audience members match your information against the background of their beliefs. Sharing your knowledge encounters the requirement of matching knowledge levels to the audience's beliefs. To share also requires avoiding rejection by their heuristics and biases. The degree of knowledge match determines neural processing in accepting, rejecting or modifying your information into audience knowledge.

Understanding messages varies and depends on the degree of overlap in belief systems, knowledge, and motivation to adjust conflicts in audience beliefs. From a social perspective, the communication of ideas is a problem that is closely related to group identity. Science depends on sharing information and communication works best when the audience comprises individuals who work in a similar field as the speaker. We also see this in how society is structured into a hierarchy of groups with variable levels of communication possible between groups. Society works as groups that develop a shared worldview. This view strongly interacts with their acceptance of what problems are and what constitutes evidence. Conflict between these views is what creates informational barriers to new beliefs. In order for society to function and change, it must establish shared goals that are visions of a common future. Goals emerge within social groups as imagined futures that offer the promise of a better life. Visions of future rewards provide both a collective motivation and the energy and sacrifice that is required to restructure society as it moves toward the goal. In politics, this is the avenue where grass roots pressure works to motivate legislative and judicial actions on new initiatives.

Why don't societies work as well as individuals in behaving optimally?

The question must always be asked as to why public policy is so hard when to each of us the appropriate direction appears obvious and rational. Instead of society being directional in its response and using the appropriate tools to head off developing problems, we seem to move from crisis to crisis without agreement on error detection and course correction. Whereas evolution supplied us the tools to be rational and to respond adaptively, we nonetheless are overcome by the complexities of societal-cultural alternatives and conflicts in developed world-views. This becomes the problem we face as disparate groups' world-views collectively produce intrusions into the worlds of others. If uncorrected, these errors can reach proportions where they impact reality in the world that we all depend upon for life support. Ultimately, society must comprehend the question posed by Jared Diamond, *"Will non-sustainable developments become halted in pleasant ways of our choice, or in unpleasant ways not of our choice?"* This is the reason environmental issues must be treated by adaptive responses from a public that wishes to have a future. Diamond's question applies to a broad set of non-virtuous directions chosen by our world's societies.

I have presented public policy as folk psychology concepts related to how the underlying models and process in functioning minds respond to new ideas. This was presented at the level of symbolic control rather than the neural background, as this best describes the functions in behavior. Knowledge is intrinsic to the individual brain and only the information that shapes these models can be shared. Public policy can only be changed as new ideas work through individuals to become a community process. An idea represents the emergence of a new model that catches the attention of conscious processes, which then can work to shape and reinforce it. Success in this process leads to development of new ideas that turn into beliefs. Societies develop their culture by individuals sharing information as ideas that develop into beliefs. This is the process that converts individual behavioral models into shared beliefs, values, and goals. Because only information can be shared, the effective packaging and presentation of information is the key to allowing interactions of the content of your models with the models in the minds of others. New ideas in systems that are complex and technical have barriers to shared information and its relevance to problem areas.

From cells to societies: biological and political "cancers."

It has been easier to share ideologically aligned information in the social, cognitive, and motivational framework of politically conservative minds. Will Rogers reflected on this in the early 1930s and it still rings true: *"Democrats never agree on anything, that's why they're Democrats. If they agreed with each other, they'd be Republicans."* Conservatives operate in tighter social groups with aligned media outlets where the same concepts in the same words are repeated. New ideas based on new data are a harder sell than traditional ideology but ignoring new ideas and information can lead to crises. We must always remember the lesson taught to us by cancer cells that cut themselves off from the communication and cooperation that is necessary to govern the decisions and responses, which will provide an adaptive future for the society as a whole. This prepares us for a morality story based on the importance of how biochemistry makes decisions to run a cell as a member of a democracy and the overwhelming importance of communication and just decision-making.

I have earlier analogized the cells of the human body with citizens in a society; they each fulfill specific roles but are in open communication that maintains favorable conditions for the group as a whole. They are quite varied in their duties, beliefs, and desires. Some are stable and conservative in their actions based on adaptations for historical reasons; others resemble liberals in being more open to innovation and change. However, they maintain open communication and equitable exchange of resources. If they function well, they make the best compromises that provide for a better future. Breakdown in communication could lead to dire consequences and this is what happens when cancers appear. Within a cell a genetic change can lead to a breakdown in communication of its normal epigenetic program. The dysfunctional systems of such a cell resist attempts to correct these errors. The cell then becomes reprogramed in gene expression, metabolism, and cell cycle. It is now dedicated to its own interests and uses all the resources it can acquire for its own growth and replication. The cells divide and their descendants all share these selfish properties. They stimulate supply of the resources they need to grow and proliferate massively. Then they send out their descendants to set up enclaves in other regions of the body not yet exposed to invaders, with the selfish desire for growth and power. This all occurs by violating the rules regulating a democratic society, breaking down open communication, and the spreading of lies. As they metastasize, these cells can collapse the system that once worked for the benefit of all, but now diverts resources to the members of the invaders. The story shows the importance of following a virtuous life whether you are a cell in a body or an individual in society. Some recent studies indicate the cancer cells act as they do because they are trying to revert back to regulatory pathways that ran their lives in simpler times when they lived as separate single cells. The studies show this as inactivating genes involved in evolution of multicellularity.

Here I have an evolutionary morality story that should carry a powerful message regarding how to run a society rationally. John Rawls 'veil of ignorance' and Karl Popper's *'Open Society'* each speak to the role of dialogue and justice in deciding who gets heard and how wisdom must be used in decisions that affect everyone. The cancer that appears in the breakdown of this democratic dialogue exemplifies current problems in governance of human society. The world needs a TP53 tribunal, since if these selfish authoritarian components of a society are not rehabilitated, their spread threatens the whole system. This makes societal cancers a responsibility of the whole community. This requires us all to have the courage to face up to authoritarians, dogma, ideology, and self-serving lies. Justice requires wisdom acquired from all the evidence and each person must have the courage to educate themselves on issues to vote for an equitable society. Democracy is a complex process and to enjoy its benefits requires its members to participate in a virtuous way. When all elements of a society play thoughtful empathic and symbiotic roles, they provide all a regulated functional existence. This is required to shape a society to cope with changing conditions so it is flexible and can even try new directions. Through 4 billion years of practice and correcting errors, evolution and learning provide us a model along with the tools to create a virtuous society.

Further Reading and References.

Gopnik, A. (2016). *The Gardener and the Carpenter.* New York: Farrar, Straus and Giroux.

Kaplan, J.T., S. I. Gimbel and S. Harris (2017). Neural correlates of maintaining one's political beliefs in the face of counterevidence. *Scientific Reports*, 6:39589.

Kanai, R., T. Feilden, C. Firth, and G. Rees. (2011). Political Orientations Are Correlated with Brain Structure in Young Adults. Current Biology 21: 677–680.

Menon, V. and L. Q. Uddin (2010). Saliency, switching, attention and control: a network model of insula function. *Brain Structure and Function* 214: 655–667.

Popper, K. (1945). *The Open Society and Its Enemies.* Princeton: Princeton University Press.

Rawls, J. (1971). *Theory of Justice.* Cambridge, MA: Harvard University Press.

CHAPTER NINE

Religion for Century 21

"And it is a strange thing that most of the feeling we call religious, most of the mystical outcrying which is one of the most prized and used and desired reactions of our species, is really the understanding and the attempt to say that man is related to the whole thing, related inextricably to all reality, known and unknowable.... It is advisable to look from the tide pool to the stars and then back to the tide pool again." John Steinbeck The Log from the Sea of Cortez (1951)

"Wonder is the feeling of the philosopher, and philosophy begins in wonder." Plato who may have been influenced by Socrates' statement: *"For wisdom begins in wonder."*

"The most important task confronting mankind is to reinvent the sacred." Scott Momaday 1992

The ultimate questions we seek to answer.

Ontology is the branch of metaphysics concerned with the nature of being. For the religiously inclined, it is a search for a supreme being and a quest for human purpose. For others, not wed to organized faiths, it more embodies a quest for ultimate explanations to basic questions such as: Why is there something rather than nothing? What is matter? And why does nature work in the first place? The concept of God, gods, spirits and imagined guiding forces provides meaning to a story in addition to the basic story of existence. The search for reasons for existence and meaning in society is driven by language and communication that requires understanding based on symbolism where shared symbols relate to explanations of causality.

Humans' curiosity for final explanations of causality has led to a continued search for hidden symbols. Since ineffable feelings are what meaning is all about, this search for godlike mechanism has taken us beyond physics, where we have shared symbols to use as metaphors in communication. The challenges of communication are shared also by consciousness, which for each of us must include our god concept for the final understanding.

Our curiosity of unanswered questions derives from some relation between emotion and the knowledge state of our neural models. These models are committed to explain the flow of information concerning our relation to the world. Meaningful models in a complex and dynamic world must incorporate a search for potential errors as well as a search for deeper understanding and meaning. The magnitude of the mismatch between our models and reality is Karl Friston's free energy, which is the Bayesian motivation to continually improve our models to match experienced reality and iteratively predict behavior for our future. Thus, this need for god or spirituality may essentially be the motivation driving our continuing adaptation to a world we wish to understand and interact with in more complex and creative ways. We seek to provide a model for this need that we can share through language, which, however, requires a shared ontology that is always it seems just beyond reach. Those who give up reaching for answers individually find a solution through membership in a reassuring group led by the unquestioned authority of their deity figure. Alternately, a more open and questioning kind of spirituality would celebrate this search for knowing by fostering the appreciation of the sense of wonder. Wonder will lead to the experience of achieving fuller participation in the miraculous story of our unfolding universe. That which is religious to one is sacred, worshiped, numinous and provides deep meaning and purpose before understanding.

Holism, Steinbeck, Tide Pools, and the Sea of Cortez.

When it comes to religious conversion, often one can point to a pivotal event in one's life that set them on that path. Mine began with the ordinary TGIF tradition at Hopkins Marine Station on some Friday in 2003. This is when Stanford biologist professor Bill Gilly had recently returned from a trip to the Sea of Cortez. Gilly was a close friend and a neurobiologist who had been doing studies on the Humboldt squid that had recently moved into that gulf. Gilly had just read *Sea of Cortez* by John Steinbeck and Ed Ricketts that documented their cruise exploring the biology of that region. He proposed rerunning the 1940s Steinbeck and Ricketts cruise with a group of colleagues including myself. I expressed enthusiasm thinking it would never happen. Somehow things came together in 2004 and a cruise on the wooden-hulled vessel *Gus D* was launched.

This is a story of a how such an experience can produce a change in perspective, philosophy, intellectual endeavor, and pedagogy and was an adventure that led to my holistic spiritual conversion. This will be a personal story, since it is only my consciousness I can look into to describe the remodeling of my mind. However, the stimulus, motivation and opportunity derived from Gilly's exposure to the philosophy of Steinbeck and Ricketts that we shared and discussed on the cruise. In addition to my conversion, this expedition would lead to Gilly's commitment to teach Stanford undergraduates in a new way— creating a course in holistic biology which included me.

A distinct challenge arises, however, when you teach a course and don't fully understand what its title means. What is holistic biology? Teaching this course with Gilly launched me on my quest for answers and what the goals of the course should be. Aristotle told us that the whole is more than the sum of the parts. Jan Smuts in 1927, coincidentally, the year of my birth, impressed by new directions in science coined the term in his book, *"Holism and Evolution"*. He emphasized that Holism should stand along with Materialism and Spiritualism as a theory of the Universe. I had previously danced around the edges of this topic through reading about complexity, chaos, fractals, ecosystems, earth systems, and multidisciplinary approaches. Now I had to get serious and organize my thoughts to create bridges from reductionism to holism and include new interactive disciplines along with science. The topics included the domains of history, law, sociology, economics, arts, poetry, spirituality, philosophy, and literature. It is always interesting to become involved in ventures that lie beyond one's level of competence. My next decade would be very busy.

Steinbeck advises us to look from the tide pools to the stars and then to look back again. He introduces the metaphor of the tide pool as a model for human society. Our lives are linked to the cosmos reaching back to the Big Bang and to all other life on our planet through evolution and ecology. We must explore the whole story to understand the meaning of our existence and the essence of spirituality. Aristotle observed that "humans by their nature are curious" and if this thought could be unraveled it not only has deep philosophical meaning but also must possess a physiological basis. Curiosity has a relationship to exploration and creativity. These activities involve agency, evolution, and the adaptive value of knowledge gained from the past to predict the future. The quest of life has been to expand exploration of the physical world in order to understand it for adaptive purposes. This quest began from the humblest explorations of organic molecular replication and proceeds through evolution into hierarchically arranged biochemical systems of agency resulting in what we recognize as life. From these early processes, emerge successive marvels such as metabolic pathways, bacterial chemotaxis and behavioral changes as seen for example in the protozoan, *Stentor* to a repetitive stimulus. They culminate in the metazoan invention of the nervous system to flexibly use the immediate past to optimize behavior for the future through learning.

The cognitive human species became social and acquired symbolic communication, which in turn led to our multigenerational cultural story. This has led humans into a dimension of curiosity unexplored by any other known life form on our planet. Humans have the unique ability to ask shared questions about the causes of regularities we observe and hypothesize explanatory stories that we can then communicate culturally. This activity creates a competition of understanding between different stories and has created the histories of diverse cultures which we still attempt to reconcile and unify. The last 50,000 years, during which humans have had language, has been a time of an ever-accelerating development of intelligence. We treat information as knowledge and seek to understand our world, ourselves and our place in the order of things. Our quest has not been without stalls and bumps in progress. Over the last 550 years, the growth in what we know and understand about the world and our place in it has expanded exponentially in the minds of scientists. Their work has created our physics. But the story of humans and life in our solar system is only a miniscule fraction of the whole saga. With upwards of 10^{22} other stars in the Universe we can only wonder what their stories are. Our tale began 14 billion years ago and recent developments in the fields of science have only just now enabled us to unravel and chronicle this greatest of stories—the story of universal creation. Our universe began about 14 billion years ago and developments in the fields of science have enabled the detective work to unravel and chronicle the greatest story that could ever be told – the story of creation from the Big Bang.

Quark Soup for the Soul

Creation is an epic success story as told from planet Earth, a pinpoint in the Universe. It begins with quark soup and carries us to the origin of life on our planet that eventually leads to the sophisticated complexity of human civilization. What we relate when we tell our story is only the most recent version that has been told and retold many times and many ways in many cultures. This version is one that has been pieced together by generations of scientists working independently in disparate fields of research. Science is a self-correcting paradigm where the parts have to fit together without being ambiguous or contradictory or fall outside the accepted rules of the game.

Religion, philosophy, and science all arise from an attempt to explain to ourselves what there is, where it came from, and what it means. Many of the previous explanations are good stories that served the purposes for which they were created. However, unless our universe is the product of a sophisticated holographic computer game of some exceptional alien intelligence where the rules can change arbitrarily, there can only be one correct story. For an inflexible organized religion to survive in the face of an increasing and diverse knowledgebase, its basic tenets must be accepted without being questioned or continually changed. Such an approach however would spell the death of science and philosophy. For both philosophy and science, the quest of their practitioners is to question authority, modify and even overthrow accepted beliefs since that has proven to be the road to both fame and progress. Organized religion treats answers that can't be questioned; philosophy treats questions that can't be answered, according to a comment I remember but can't source. Scientists answer some questions so they can ask new questions that are more fundamental. However, progress is dependent on correcting errors in performance, thus the critical importance of getting it more right.... or less wrong if we are to be pessimistic.

From Quarks to a Quest for Understanding.

As I suggest, the story goes back nearly 14 billion years and research efforts of physicists and cosmologists have penetrated back to just after the separation of gravity from the other forces at 10^{-43} seconds after the beginning. A more rigorous understanding begins at 10^{-36} seconds when the strong nuclear force separated from the electroweak force. From this event to 10^{-32} seconds, there was an expansion of space by a factor of 10^{30} expanding the nascent universe from proton size to the size of a grapefruit. This was an intensely hot quark soup; with further cooling, by 10^{-6} seconds all of our present forces and particles were present. By 10 seconds, we had nuclei of hydrogen and helium. This is a beautiful story derived from the standard model of particle physics unfolding from the Big Bang in the cosmological evolution of the universe. From some compressed pinpoint of an intensely hot fusion of all that would be matter and energy, 100 billion galaxies unfurled. On the average, each contains 100 billion stars. Many of these stars have life hospitable planetary systems, simulating our curiosity. Our 4.54 billion year old planet has provided the environment that became the crucible for the only life we currently know of in the Universe. Earth provides a platform from which we examine the Universe's history and development.

Life on Earth has a history that begins 4 billion years ago as our planet cooled, and biochemical evolution of complex carbon-based molecules proceeded. In a couple of hundred million years this would produce the microbial cell, LUCA, the last ancestor common to all Earth's life. After 2 billion years some of LUCA's descendants would symbiotically fuse to produce our chimeric eukaryotic nucleated cells. These are the cells that invented sex and evolved multiple ways to assemble into larger multicellular forms of life. Nucleated life diversified into several dozen kingdoms of organisms, a few would achieve the ultimate in complex multicellularity as our familiar fungi, plants, and animals. This is of course the story of evolution, an emergent system that grew out of developing chemical complexity. In a sense, evolution is a process of self-replicating systems that leverage off a history of past successes, hoping to predict the future. Life can do this because it is self-correcting and eliminates mistakes in order to survive.

Control theory informs us that in order to interact with another system we must have a model of that system. Life began with the agency of behavioral biochemistry where evolution crafted chemistry into systems linked by information in its structural models to detect and interact with other systems.

These behavioral interactions provided the resources necessary to maintain the structure and replication of the models. These models introduced knowledge, information, purpose, and meaning into the ongoing physics of our world. They thus became biochemical agents with the ability to assess conditions and respond adaptively. Some of these agents began to cooperate as linked models in biochemical pathways, which would lead to components for the evolution of the first cell perhaps 3.8 billion years ago. The time from the beginning for our ancestral cell to appear was 10 billion years. All life on our planet is linked to this ancestor including our animal line that gave us a nervous system around 550 million years ago. In animals, nervous systems comprise cellular pathways that conduct information from receptors to a set of cells that predict a behavioral response. This involves the use of a growing, flexible, and correctable memory that contains the adaptive meaning of input it has received. These systems have become immensely complex during animal evolution and provide humans with our mental capabilities. Our language, culture, stories, and marvels of civilizations along with its attendant problems emerged from this innovation. Our story of creation has emerged in the last few hundred years. This leads us to wonder about all the other stories of creativity. How can quarks produce the quest to understand how all this came to be and why does it have meaning?

The Importance of a Sense of Wonder.

I have been intrigued by the sense of wonder. Our behavior is based on the prediction that we will experience the future world as we have known it in our past. When the unexpected occurs, we depend on creativity, which is the exploration of new challenges and opportunities that are presented unexpectedly. The unexpected is associated with feelings of surprise and a sense of wonder, which provides motivation to incorporate this new knowledge into our world. One of my earliest childhood memories is on a vacant lot where my older cousin used his pocket knife to pry up a round bit of dirt and exposed a silk-lined tunnel; a huge black spider came crawling up, grasped the trapdoor and struggled to pull it closed. This may not have been my first such experience, but it was followed by the early years of my life being involved with revelations from the world of nature. Such early experiences were perhaps interacting formatively with the inner essence of my mind, which would guide my life into a career involved with nature.

These stories of structuring the mind are too complex and hidden to yield to easy explanations, but the path to understanding is hinted at by the studies of UC Berkeley Psychology Professor and author Allison Gopnik. She believes that babies and young children act as little scientists that create and explain their world through their observations, forming hypotheses, testing these, and enjoying the rewards of successful interactions. Young humans proceed with delight while unaware of the process, which is controlled and motivated by their unconscious selves and experienced in their consciousness. The sense of wonder motivates babies and us to investigate. Gopnik also wrote a paper for a philosophical journal, *"The Scientist as Child."* Here are some thoughtful words from the paper's Abstract that make us wonder why some abandon the ability as they mature?

> *"...powerful and flexible cognitive devices that were designed by evolution to facilitate learning in young children. Both science and cognitive development involve abstract, coherent systems of entities and rules, theories. In both cases, theories provide predictions, explanations, and interpretations. In both, theories change in characteristic ways in response to counterevidence."*

I present some thoughtful quotes on the sense of wonder below. These quotes express something about the cognitive structure and motivations of the minds of four students of natural history who incorporated their love of nature into the productivity of their careers.

> *All the phenomena of nature need to be seen from the point of view of wonder and awe. Walden,* Thoreau 1852

> *It is interesting to contemplate a tangled bank, clothed with many plants of many kinds, with birds singing on the bushes, with various insects flitting about, and with worms crawling through the damp earth, and to reflect that these elaborately constructed forms, so different from each other, and dependent upon each other in so complex a manner, have all been produced by laws acting around us. On the Origin of Species,* Darwin 1859

A child's world is fresh and new and beautiful, full of wonder and excitement. It is our misfortune that for most of us that clear-eyed vision, that true instinct for what is beautiful and awe inspiring, is dimmed and even lost before we reach adulthood. The Sense of Wonder, Carson 1956

Ecological wisdom involves the ability to see the beauty in nature and to integrate it into the patterns and processes studied by ecologist. Dayton and Sala *The Sense of Wonder*, Scientia Marina, 2001

Thoreau immersed his life, thoughts, and philosophy deeply in nature and used species and communities as analogies for moral and social values. He viewed natural systems as holistic and dynamic and in a perpetual state of becoming. Thoreau, with later support from Aldo Leopold and John Muir, introduced our country to the environmental ethic. The drumbeat he marched to is still inspirational to those who would lead a thoughtful, moral, and self-examined life.

Darwin provides the ultimate example of the holistic synthetic naturalist. Peter Grant, is a distinguished evolutionary biologist and naturalist emeritus at Princeton University. In an address to the American Society of Naturalists he defined a naturalist as a person who asks questions of nature and answers them by using every means possible. Darwin transformed from a collector of beetles to one of the world's most influential scientists by adherence to this dictum. He contemplated a career as an English country parson, indulging himself in the local natural history. His plans were sidetracked by an opportunity for a long sea voyage. He began that voyage with an open mind [and a weak stomach] and a broad background in the natural sciences. During its course, he experienced an intense exposure to biogeographic patterns of biodiversity. As a result of these experiences, he was driven to ask questions about the mutability of species. The rest of the story is history, to the extent that, as the geneticist and evolutionary biologist Theodosius Dobzhansky has reminded us, nothing in the world of biology makes sense except in the light of evolution.

Under the Sea Wind, The Sea Around Us, The Edge of the Sea, Silent Spring and *The Sense of Wonder* are a set of books by Rachel Carson that translate the spirit of a naturalist to anyone who can read and also to children who must be read to. Rachel Carson's prose has a lyric style, which paints pictures in our minds as she weaves strands of science and nature into narrative. Her stated approach to writing was synthetically holistic. That she created from her unconscious self as we know from Ellen Levine's book, *Up Close: Rachel Carson*....

> "*know the subject intimately... let it fill the mind.... At some point the subject takes command and the true act of creation begins... the writer should be still and listen to what the subject has to tell him.*"

She is best known for intuitively recognizing and responding to the reality that human activities could produce unintended and catastrophic changes in nature on a global scale. Carson made this point clearly and, thereby, shocked the world with her book *Silent Spring*. If only more humans had incorporated her Sense of Wonder into their relationship with the natural world, *Silent Spring* would not have been necessary.

Paul Dayton's quote is taken from a journal article where, with his provocative philosophical approach, he shares his thoughts on the importance of natural history in understanding the complex dynamic systems of nature.

Dayton emphasizes, "*Creative ecology is based on a deep sensitivity to natural patterns and processes. Naturalists have the ability to synthesize perceptions of nature into reasonable hypotheses about processes that cause the patterns, and then shift into the relatively simple scientific technology of testing hypotheses such that they contribute to a more general understanding of nature. In this sense, good natural history is fundamental to ecological science.*"

Studies in ecology and on environmental issues continually become more multidisciplinary and require team efforts, but until we can achieve knowledge transfer between minds, melding the creative synthesis of disparate sets of perceptions into explanations of system process will likely remain the product of a naturalist mind.

Sense of wonder recurs throughout Thoreau, Darwin, Carson, and Dayton, and speaks to a relationship of humans to nature on their individual subjective emotional states, rather than just on utilitarian levels. Our society continually alters our shared environment with human-produced effects of magnitudes that seemed to make the problems insoluble. Einstein warned us, "*No problem can be solved from the same level of consciousness that created it.*" Some thoughtful individuals have sought answers in a deeper awareness outside of the societal activities that produced these problems and the efforts of organized science to try to solve them. Ed Ricketts, in the late 1940s, despaired of solving the depleted sardine fishery by educating the fishermen and felt that perhaps a Sunday sermon at church could better convince the fishing industry of its salvation through conservation.

Robert Nadeau in his book, *The Wealth of Nature,* despairs of the ability of economically based decisions or the current procedure of science to solve looming environmental problems. He feels that blending a type of value-based spiritualism into science can make it emotionally interactive with society to solve environmental problems. E. O. Wilson in *Consilience* expressed a hope to achieve solutions to these types of problems by a fusion of the many disparate areas of knowledge, which would result in a balance of science, values, and society to reduce conflicts and enable solutions. The concept of consilience came from the English polymath and originator of the word "scientist," William Whewell (1794-1866), who proposed, *"The Consilience of Inductions takes place when an Induction, obtained from one class of facts, coincides with an Induction obtained from another different class."* However, Wilson does not suggest how to accomplish this fusion of concepts in those minds that achieve decision-making status. We have suggested the importance of the diversity of experiences encoded in our autobiographical memories, which can then be metaphorically linked in our default mode network to approach new problems successfully.

The Canadian geneticist and environmentalist David Suzuki (b. 1936) drew inspiration for solutions to society's environmental issues from aboriginal people who literally believed that they were part of nature. In his book, *The Sacred Balance,* he incorporates philosophy and science with theology and spiritual appreciation of nature by humans as a way to rediscover our fundamental relationship to nature. Suzuki proposes that while there is no 'balance of nature' there must be a balance in humans' interaction with it. Balance is what is meant by sustainability. The American scholar Edith Cobb's (1895–1977) book, *The Ecology of Imagination in Childhood,* documents that in downloading formative early memories of creative adults, there is a strong sharing of an *"early awareness of some primary relatedness to earth and universe".* She develops a story of imaginative childhood play in nature as being what forms the mental machinery, which later in adulthood enables creative work. This idea parallels the discoveries of Alison Gopnik, i.e., that those who retain childhood memories of imaginative play are more likely to do creative work as adults.

We need to "Reinvent the Sacred."

Early in my quest to see how understanding of cognition and the control of behavior could help in reaching solutions to our many pressing problems, I was reading on the origin of life and ran across references to the work of Stuart Kauffman. I acquired his book, *Reinventing the Sacred,* and found it to be about much more than chemical evolution. The title came from his participation in a conference of four people confronting the question of the most important issue facing mankind. These are his words from the notes to the first chapter. "*. . . a magical mountain of a man, Scott Momaday, Pulitzer Prize–winning Kiowa poet. Scott, perhaps six feet seven inches, some 270 pounds, bass voice, fixed us in his gaze and said,*

> "*The most important task confronting mankind is to reinvent the sacred." I was stunned. Trained as a doctor and scientist, even with a background in philosophy, it was beyond my ken to use the word sacred. The topic was outside the pale of my view of informed conversation. And I was instantaneously convinced that Scott was right.*" The group's position statement at the end of the meeting was fascinating. These are in the notes to the first chapter: "*We wrote, roughly, that a global civilization was emerging, that we were entering its early "heroic" age, when a new transnational mythic structure could be created to sustain and guide that global civilization, that we could expect fear and fundamentalist retrenchment as older civilizations were challenged, and that reinventing the sacred was part of easing the emergence of a global civilization.*"

This book was inspirational and helpful in many ways to me and it carried the spiritual message of the sense of wonder to be found as "*the natural creativity of the universe.*" Kauffman makes the case that life has emergent properties and while these properties do not violate the rules of physics, they nonetheless are not reducible to these rules. This creativity of life then creates wonders that we participate in and the tenets of a sacred belief system must engage rational and moral responsibility. I treasure his statement,

> "*If we reinvent the sacred to mean the wonder of the creativity in the universe, biosphere, human history, and culture, are we*

not inevitably invited to honor all of life and the planet that sustains it?"

In a way, this sense of wonder, this reinventing/redefining of the sacred, can be compared to the spiritual experience or sense of divinity that has played a big role in the commitment of people to organized religion. Perhaps this phrase describes the emotional systems' motivation that works below awareness in the default mode. Emotional bonding with nature enables a level of spiritual interaction, which inspires holistic understanding and caring interactions. From the perspective of our environmental concerns, can we teach or evoke this spiritual sense of wonder in the broader public? Or is it more like love that develops in an individual where it is personal and experiential, but incredibly diverse in what it means for each person? Daniel Dennett claims that even for those in the same religious sect, everyone's concept of God is different and individualistic. The problem we face in relating our human population to earth systems awareness is perhaps not so much a problem of education as it is a problem in the reconciliation of every individual's emotional landscape.

From a cognitive perspective, is the sense of wonder just one of the facets of our creative soul similar to that which empowers the production and appreciation of art, music, and poetry? For those who relate to the natural world, our sense of wonder creates a conscious awareness of experiences with nature, which is colored with emotional meaning. This same effect is what gives religion its power when people portray their lives as enactive with and benefiting from a biblical scenario. There are many stories of people with personal troubles who have found salvation through conversion to religion. How can a similar conversion be made to happen that fosters a person's desire to protect our troubled environment? Any conversion must involve redirecting focus of the emotional and motivational meaning and interpretation of pertinent knowledge that shapes a person's current world-view. This need brings us to the problem of how emotion and motivation influence the magnitude of probabilities we assign to our mental hypotheses and thus their reevaluation in the light of new data. This is an important cognitive issue because what religious conversion does is to selectively alter attitudes and emotional likelihoods pertaining to certain areas of our knowledge where there was probably already some troubling cognitive dissonance. However, conversion to the new doctrine also introduces conflicts with other knowledge or hypotheses which now require adjusting to new beliefs.

Sustainability: A philosophy of adaptive ecosystem management, by Bryan Norton provides some thoughts on environmental conversions taken from the views of Thoreau and Leopold:

> *"a holistic jump is necessary for environmental revelation. It is both a revelation and spiritual rebirth..." "the new consciousness cannot be expressed, much less justified, in the immature consciousness. The transformation is in this sense nondeterministic and requires an intuitive spark as much as observation and logic."*

In *Thinking Like a Mountain*, Aldo Leopold documents that his environmentalist conversion was inspired by gazing into the eyes of a dying wolf, with his concluding analysis; *"Perhaps this is the hidden meaning in the howl of the wolf, long known among mountains, but seldom perceived among men"*. He had perceived that mountains change on a different time scale than the lives of humans. This pays homage to Thoreau's classic thought *"In wildness is the salvation of the world"*. Cathedrals and religious art play a large role in conveying the sense of awe, wonder, and meaning that reinforces the spiritual experience that bonds a person to their sacred values. Nature plays this role for me, and I'm sure others, especially as it becomes translated into process and meaning. A sunrise in Baja with a school of dolphins playing in the shimmering Gulf creates a numinous experience in my mind.

Life is a rational game with dynamic rules. Hypotheses established the beginning of life and originated the rules for the game of life. These hypotheses and rules began as early predictions of how to acquire resources. They arise as minimally tested Bayesian prior beliefs that are tested against further evidence as they are repeated, corrected, and refined. These hypotheses or beliefs are models of behavior. They are dynamic rules of conduct, which evolve, proliferate, and diversify for different goals. These rules are assembled and networked in an individual and they both define and interact with its environment. Life is an explanation of adaptive intervention on causality, which is ongoing in the world. Life is a behavioral system, which maintains itself by adaptive model selection, assembly, and prioritization.

Does something like the sense of wonder drive the game of life? Exploration in a complex world presents challenges and opportunities. If it is

Bayesian induction making up the rules, then observation would evoke priors and test for an explanatory fit to the surprise of the observation. Wonder driving the fit of the explanation establishes new rules that are behavior. These rules are based on neural circuits, which interact with the new pattern. The complexity of life's solutions to these wonders of the world is manifested by evolved biological diversity. For the last four billion years, untold numbers of behaviors have interacted with our world and we have emerged as a remarkable product. It has been claimed that humans are the most complex entities in the known universe. This is because of the human brain, which we use to create our sense of wonder and stories about it all. I wonder if the sense of wonder is a high positive emotional priority of the free energy associated with new input that can tell me a story I want to hear?

My Spiritual Conversion to a Pantheism: A Summing Up.

My quest of course leads to the story of my spiritual conversion. Most of what I know and what I am, I learned from the better part of a century by interacting with friends, students, and colleagues. Therefore, they deserve some of the blame for my religious conversion, especially because the holistic biology course was the reason it happened. Zealous converts feel compelled to share their path to enlightenment and thus my story will deal with my conversion. Every convert has a personal history of a continuing degradation of their soul and spirit by the trials and tribulations of living in a world they cannot control. After retirement from MBARI, my third retirement was from environmental television programming in 2002. This retirement gave me more time to read the newspaper, listen to the news, and worry about how the world was sliding down the slope of catastrophe and how humanity was totally unable to convert the knowledge of science into societal wisdom. Could our earth have worked four and a half billion years to produce intelligent life, only to have it squish itself out of existence in a few thousand years?

I tried displacement activities such as: 1) making nature movies and drinking old- fashions; 2) reading and thinking about cosmology and phylogeography and drinking Shiraz; and 3) ultimately going on a cruise to the Sea of Cortez and immersing myself in holistic philosophical issues, experiencing historical ecology and celebrating the invasion of the semi-giant squid while drinking fermented malt beverages aboard the *Gus D*. However, there was always the return to the reality of television, radio, and newspapers and my soul and spirit would again go into a tailspin.

I didn't work at Stanford anymore but through a complex set of circumstances I wound up involved in the first course of holistic biology and became responsible for playing the role Ed Ricketts would have played as a naturalist and philosopher in teaching the course. Areas of Ed's perspective where I was deficient required my preparation in poetry, baroque music, and getting into Eastern philosophy and religion. This resulted in my introduction to the philosophy of Taoism and the deep significance that could be found within 2500 year-old documents. I became intrigued by the writings in *Tao Te Ching* and how it regarded man as part of nature and also spoke to curbing the excesses of rulers who would disrupt the natural balance and flow of the fundamental process. Since I began reading during the turbulent years of the first decade of the new millennium and Taoism spoke to curbing the excesses of rulers, I was struck with how these views resonated with problems we face today.

I didn't have a spiritual conversion to Taoism because it didn't readily translate to our contemporary world and it is more like a philosophy than a religion. For some time, I had been bumbling along with my soul and behavior generally guided by humanism, which is an atheistic philosophy that stresses the individual assuming responsibility for their ethical choices...... But remember, I had this ongoing escalating erosion of soul and spirit that required some kind of positive reinforcement to reverse it. Moreover, a number of sources in my reading on solutions to environmental issues had concluded that some type of religion was perhaps a salvation for our environmental crisis. This also troubled me as I began my education in the pragmatic world of engineering and then as a biologist became evermore intellectually secular and removed from mysticism. One might say I was still a deeply troubled person, but when I began Googling around in Tao stuff, I hit on a URL that introduced me to Pantheism.

Reading on the tenets of Pantheism provided concepts that I found to be rational. Tenets include a physicalism, where humans are a part of nature and value thoughtful interactions with our world. Life emerged from the physical world and the body and mind emerged from matter by natural process. Their beliefs stress reality and the use of our senses in understanding through evidence available to us rather than from authority. Tenets support a world community based on freedom, democracy, justice, separation of religion from state and world peace. Our actions, ideas and their consequences are what lives on as our participation in the future.

I have come to believe that the origin of life as behavior and cognition is the origin of the rudiments of meaning and values, which can later emerge in spiritualism and the sense of the sacred. As I stressed in the previous chapter, social behavior in groups of agents requires virtues for morality. Since life occupies a complex dynamic world, these virtues reside in systems that have evolved, but maintained an empathic quality, tempering competition. Human societies have proposed many cultural approaches to explain where we came from, what it all means, and how we should behave. This has introduced problems. Pantheism is a spiritual view that spreads the concept of the sacred and the concept of divinity broadly across the Universe. Linkage of humans in society and humans with nature is reality and participation and understanding in the process is my divinity. Astronomer Carolyn Porco expresses this beautifully,

> *"At the heart of every scientific inquiry is a deep spiritual quest — to grasp, to know, to feel connected through an understanding of the secrets of the natural world, to have a sense of one's part in the greater whole."*

The whole essay can be accessed at, *The Greatest Story Ever Told, Edge, The Most Dangerous Idea*, 2006. The World Pantheist Movement provides a credo that describes its beliefs, and here I quote from the first of these. *"We revere and celebrate the Universe as the totality of being, past, present and future. It is self-organizing, ever-evolving and inexhaustibly diverse."* Reading on and from pantheists provided a religious philosophy that fits my life, beliefs, morals, science, and teaching. In reading quotes on the doctrine, I discovered my inclusion in a large diverse group of declared pantheists whom I admired, including Einstein. To paraphrase Will Rogers' comment on Democrats; I don't belong to any organized religion, I'm a Pantheist.

I can't really say I was converted or saved since I discovered that this was the direction my "religious" philosophy had been developing as a core set of beliefs throughout my life; thus, I didn't have to learn the gospel. Though it had not been easy it had been fun; while reading, thinking, loving, communing with nature, and talking with people, I had become a pantheist without even knowing it. On my discovery of being a pantheist, it motivated me to holistically convert my separate experiences into a related theme with deeper meaning. It was also rewarding since I had been a practicing minister since the turn of the millennium but could only perform nondenominational weddings since I had no religion. My first wedding as a pantheist minister was a holistic ceremony for the union of my colleagues and friends, Susan and Gilly, who had opened the path to my salvation through Holistic Biology. This conversion also opened up whole new worlds for me as a teacher since my lectures can have the deep meaning of sermons.

However, because of my deeply held beliefs I can neither convert you nor save you. I can't convert you because Pantheism is a state of the mind and only you are responsible for that. It is possible that my views could help you but only you can do the mental work necessary to achieve your internally consistent set of beliefs. I also feel that your own discoveries have deeper significance and meaning than beliefs you acquire from others. One could follow the suggested method of Thomas Nagel in *The View From Nowhere* where he says you must both understand yourself from a subjective perspective and then understand your world objectively from the standpoint of others, which includes your relation to it. I can't save you because there is nothing to save. Everything is part of this universe and stays part of it and gets recycled in it.

This of course brings us to the big question of the soul. If we read the texts we will find Pantheism is quite diverse in its positions on what the soul is and its destiny. Some early holistic thoughts on the soul were written around 250 AD by a philosopher in Rome, Plotinus: *"This universe is a single living being, embracing all living things within it, and possessing a single soul that permeates all its parts.... Soul enlivens all things with its whole self and all Soul is present everywhere..."* I believe each of us is entitled to a part of this as a personal soul that begins within us in a simple state and is nourished by the interaction of the mind with the world around us. The soul matures and becomes rich with our experiences, thoughts, and our social lives to produce the totality of our mind. It is the ever-changing matrix of our knowledge, talents, creativity, loves, fears, and our character and personality. It is our story and we acquire it in bits and pieces throughout our lives by interacting with our environment but, importantly, also with the minds and souls of others. The soul can be shared without diminishing it, so that during and after our lives parts of our soul can join with the souls of others and also be spread through the universe through communion with the natural world. Thus, it will live after us and thus the quality and meaning of our lives and our ability to share will determine the ultimate value and existence of our now chimeric soul within the universal soul.

Dogma is not permitted in Pantheism since its tenets are dependent on the structure and function of the universe and answers to questions are provisional and based on our continuing search for what is right. My current beliefs align me with the informal group of evolutionary, scientific pantheists but lately I have also been contemplating a holistic cognitive subdivision.

Further reading and References.

Carson. R. (1965). *The Sense of Wonder.* New York: Harper Row.

Dayton. P.K. and E. Sala. (2001) Natural History: The Sense of Wonder, Creativity and progress in ecology. *Scientia Marina*, 65 (Suppl. 2): 199-206.

Darwin, Charles. (1859). *On the Origin of Species: By Means of Natural Selection.* London: John Murray.

Gopnik, A. (1996). The Scientist as Child. *Philosophy of Science.* 63: 485-514.

Gopnik, A. (1998). *The Philosophical Baby: What Children's Minds Tell Us About Truth, Love, and the Meaning of Life.* New York: Farrar, Straus and Giroux.

Harrison, Paul (2013). *Elements of Pantheism: A Spirituality of Nature and the Universe.* New York: Scaftesbury, Dorsett.

Kauffman, S.A. (2008). *Reinventing the Sacred: A New View of Science, Reason and Religion.* New York: Basic Books.

Nagel, T. (1986). *The View From Nowhere.* Oxford: Oxford University Press.

Norton, B. G. (2005). *Sustainability: A Philosophy of Adaptive Ecosystem Management.* Chicago: University of Chicago Press.

Porco, C. *The Greatest Story Ever Told,* Edge essay in: Brockman, John (2006). *What Is Your Dangerous Idea?* (Edge Question Series) New York: Harper Perennial.

Thoreau, Henry David. (1854). *Walden*; or, *Life in the Woods.* Boston: Ticknor and Fields.

CHAPTER 10

Behave Yourself: The Conscious Self as Mentor

We've come a long way from our origin as stardust.

Writing a book has been a spiritual experience. It has been a search for understanding the mysteries that enshroud the essence of life. Life appeared as matter that could behave and improve its behavior by practice and by correcting errors. In emerging from the physical world, life forms evolved the ability to choose their own goals. In the words of the cosmic philosopher Carl Sagan, *"We are stardust that can choose our destiny."* From the Big Bang that would create our universe, small simple elements of hydrogen and helium formed clumps that warped spacetime into a dynamic spacescape, producing galaxies illuminated by nuclear fusion in their stars. Some stars, when they consumed their fuel, collapsed under massive gravitation to forge the heavier more complex elements necessary for life. These collapsed stars exploded as supernovae and created potential Elysian nebulae of stardust that were the nursery for more stars. These new stars were now supplied with carbon, nitrogen, and water along with other elements required for life. It was one of the hundred billion of these stars in our Milky Way galaxy, the one we call our Sun, that was encircled by the Goldilocks planet we call Earth, the planet that would be the breeding ground for life. Life appeared nearly four billion years ago and, billions of years after that, we appeared as a species with ultimate curiosity. We began to speak, develop civilizations and culture and, about five hundred years ago we created experimental science. In the short time span during which our science has developed, we have created many dimensions of understanding about how things work and can be used to our advantage. That even includes going to the moon while creating this story of our ancient history in the synopsis above.

We look for answers to questions of where we came from, what causes events, and what it all means, using religion, philosophy, and science. We have populated the Earth with societies filled with their own complexity, diversity, and contingency. Through unbounded population growth we produced cultures, religions, and governance with ideologies that challenge equity, justice, and rationality for others. This threatens our species, and life in general, from moving into a better future. Stuart Kauffman pointed out that at this stage of complexity in human society it was time to establish some sacred principles that could unify us and guide us toward our goals. He selected creativity and the discovery of new levels of understanding as the most important spiritual guides, emphasizing that science and philosophy must interact with humanities, culture, governance, aesthetics, and values. E. O. Wilson in *Consilience* treats fusion of knowledge, providing examples from the perspective of history. He reflects on how many areas of knowledge and beliefs can gain depth, breadth, and new insights from interdisciplinary approaches. He was optimistic, but even in 1998 the rate of growth of knowledge posed an insurmountable Red Queen problem of keeping up, let alone recovery and fusion from diverse archives. These efforts will remain limited and exist only in the minds of sets of experts. Though coping with complexity is difficult, expert sources must be respected rather than ignored. What is necessary for use of their product is a reverent belief in the depth of its relevance to not only our species but to the fabric of life on Earth. This belief can emerge as a shared spiritualism providing guidance on prediction of pathways to a better future. This was why religions and spiritualism were devised by humanity. Our problems are all behavioral and to get further down the road we must improve our behavior. Life as behavior is the central issue I have treated in my book, using the perspective of a naturalist.

We need a sense of wonder.

I am an academic and nearly seventy years of my life has been associated with learning, teaching, and research activities at universities. My interests were diverse but were focused around natural history. Natural history explores diversity, appreciates the wonders of the natural world, and confronts questions of how it works. I shared my natural history interests with students and found that, for many, nature provided a motivational learning experience. This was especially true when they were captivated by new experiences and felt a sense of wonder. This often led to individual explorations of topics of interest and independent observations that led in turn to questions of their own. The sense of wonder occurs in minds that are open to exploration and discovery to expand horizons of knowledge. Curiosity, spurred by the surprise of new discoveries, can be motivational. This is the driving force of innovation, science, and philosophy. It is present in babies and young children and enables them to create their view of their world and how things work. In this age of extreme change, constantly learning how things really work has critical importance. The curious, exploring, and open mind should be nourished rather than quenched. I wonder what the process is that associates the surprise of experience with the motivation to know?

Gallup conducted a poll of university graduates to discover what correlated with how engaged the respondents were in their work, as well as their community, physical, financial, and social well-being. Researchers found that six elements of an undergraduate experience had significant effects on a student's post-graduation success: A professor who made them excited to learn; a professor who cared about them as individuals; a mentor who pushed students to reach their goals; working on a long-term project; completing a job or internship related to classroom lessons; and being engaged in extracurricular activities and groups. Graduates who said they experienced all six elements were three times as likely to be engaged at work, compared with their peers who say they missed out. Just 3% of students experienced all six. How do students and mentors get together? Since education is life; how do you get mentors outside of school? How do we generate effective contact with the wonders of Nature, now that much of our educational mind shaping derives from TV and social media as chosen sources?

Mentors as coaches for cognitive functionality.

When I take time off and watch the Warriors play, I am always amazed at how they perform, and I often wonder what the Coach, Steve Kerr, has to do with it? This is perhaps because I came from UCLA where one of the greatest coaches ever (John Wooden) turned out a variety of remarkable teams using players with very different abilities. Later in their lives he is described as their important mentor for life. As we watch an event, we all become coaches and engage in counterfactual running of the team and players in what should have happened. The coach doesn't play the game, but good coaches play a huge role in the performance of those who do. What is the secret of their success and what if we had a coach that could help us with more than just physical performance?

We all **do** have a coach; that is the role of our consciousness. We feel that it runs our performance. It would seem that the consciousnesses of different individuals vary in their success in producing functionality and virtuous behavior. I began my book when I found out that our decisions and perceptions begin in our unconscious a half second or more before they appear to our conscious awareness. The study also reported that we could intervene on action as observed to play out. We observe replays and are critics of our behavior that begins in our unconscious minds. My book is an approach at establishing our recognition of this critic, and its role in coaching, teaching, directing and motivating our consciousness on how and why we behave.

Coaching our consciousness.

What I wonder is: Can you teach the coaching ability of consciousness, or is it like love, something that you have to learn on your own? There is a mutual interaction that results in loving; is this what a mentor does? What our society needs is a set of coaches for the individual's mental coach to produce winning players for their lives and as members of society. We come equipped to do this for ourselves with the tools selected for success by four billion years of evolution. Our minds are the most remarkable entities in the world, but we must work to use them well. This ability does not come naturally—it takes concerted mental effort.

I was surprised and motivated to understand how better to train my mind when I heard that my conscious perceptions and decisions had a delay of a half second before I was aware of them. My surprise was further amplified when I learned of the massive discrepancy between our conscious and unconscious processing rates. Our conscious processing of information is restricted to about 40 bits/sec while our unconscious brain works at a rate of eleven million bits each second. Below is a quote that reflects on how little of our neural processing is available in our conscious display and what the consequences of that disparity are.

> "The culture and civilization of consciousness has celebrated huge triumphs, but it also creates huge problems. The more power consciousness has over existence, the greater the problem of its paucity of information becomes. Civilization fills people with a sense of otherness and contradiction, which leads to the same kind of insanity we find in dictators surrounded by yes men." Tor Nørretranders... *The User Illusion*

Dictators can't stand to be corrected or criticized and thus surround themselves with yes men. The unconscious runs our lives by making decisions; the conscious then can act as a critic to question beliefs and correct errors. When you are not open to questioning and learning you are fostering an authoritarian dictator to run your life. This also limits your access to evidence about how the world really works. This applies to both factual knowledge and the views and feelings of others. While it feels good to think you are right, it is very difficult to be right in complex dynamic situations requiring adaptive actions.

Why do so many allow their conscious editor to support a dictatorial unconscious rather than motivating their minds as curious students do, and explore the complexity that is their lives? As civilization becomes ever more complex, open, and dynamic, the challenge to get it right becomes greater. We used to be serfs to authoritarian control because we had no access to information about how the world works. Science and the printing press fostered the Enlightenment and provided the opportunity to choose our destinies by expanding access to formerly sequestered information.

Why now, when information has never been more plentiful and accessible, are societies moving again to become the subjects of authoritarian rule? In species that lack consciousness and language, adaptive decisions are instinctual. They respond in order to survive. When our programs are well resolved and adaptive, they can run with minimal attention; however, when confronting novel or complex challenges we must solve a problem. In humans this calls for a decision on what to do next and the use of information to make the right decision. Decisions may be quick and determined intuitively or involve conscious processing, which heavily involves unconscious pathways. Humans introspectively and/or in communication with others work to justify their decisions in their consciousness. Which decisions are right and wrong reverberate through society at all levels, but our knowledge about how decisions are really made is superficial.

A rational agent is assumed to have relevant information and a capacity to assess probabilities of events. It is supposed to compute costs and benefits and select the best choice of action. Whether this choice is optimal depends on the adequacy of information. Herbert Simon pointed out that optimal decisions can only be made in simple situations and introduced the term *satisficing* as an alternative to *optimality* for decisions in complex situations. In the real world of dynamic complexity, decisions become subject to limitations by time, effort, resources and computational tractability. These limitations call for a satisfactory decision that would suffice (satisficing), albeit not in principle optimal. Then, through monitoring and error correction, decisions that are initially sub-par become progressively adaptive. An example of the importance of the satisficing approach can be seen in the arena of developing environmental legislation. The cry by industry and conservative politicians for delay in implementing environmental regulations until we have "better science" and more studies allows pollution to keep occurring. This political posture must give way to true cost/benefit reality in which laws are passed that "do the best they can," given what we know about how to reduce or solve an environmental problem.

Simon also recognized that, in addition to cognitive and situational constraints, the role of emotion must also be specified. The motivation to solve problems or avoid them derives from emotions and values. As Chris Mooney has shown, the prominent roles of fear and hope are the value differences between conservatives' and liberals' responses to change, respectively. This emotional stance determines openness to evidence and tolerance of uncertainty in viewing the future. Now with so many decisions facing society, there is the conservative choice of simplifying the world so you can claim optimality of old models that no longer fit the world; or, as a liberal, resort to the effort required for probabilistic satisficing through Bayesian optimization of all the evidence. This can only be done through inspiring minds to use hope and foster belief in visions of a better future to work towards.

The human consciousness views decisions and only guesses at the reasons used by the myriad neural networks of the unconscious brain in reaching them. This hidden processing requires a supply of evidence in order to correct errors. What is right is also influenced by our values and emotions, which have been previously built into the system by appropriate coaching. This influences the choice and use of information that is used to shape the goal of the belief model. Emotions and values are used to make decisions but are also used to justify how it feels in your conscious story. That is why being inundated by ideological yes men is destructive to normative rationality. Ideology creates arbitrary worlds where you may be happy while being wrong. Beware of the expert opinion of authorities who are so committed to their goals that they refuse to consider evidence that runs contrary to their beliefs. Look instead for those who are willing and able to acknowledge new evidence and alternate hypotheses and explore new ideas, even if some ideas go against the grain of their philosophy.

Consciousness as critic.

We should regard consciousness as a critic that is responsible for who we want to be. If your consciousness also acts as a mentor, coach, director, teacher, and motivator, it can produce a high quality product. The list of productive educational elements discovered in the Gallop Poll mentioned above matches those that act in the mind's executive function. Mentors become role models we choose to imitate, and professors tend to be successful and appreciated by students because they have well developed executive function. The executive function is empathic and fosters the professor's interest in the problems and questions of the student. The development of the executive function is the key to a balanced successful life. Invoking this function in problem solving involves having perspectives of hope and paths to a better future rather than the fear that evokes avoidance.

What I wonder is: Can the coaching ability of consciousness be taught—or must this ability be learned on one's own? This question prompts a consideration of the parallels between development of love and mentoring the coaching skill of the consciousness. There is a reciprocal interaction that facilitates the motivation for loving; is this what a mentor does? Perhaps this involves the sense of wonder as triggering an interactive resonance and romance with ideas. Multi-agent systems require sharing of values and motivation to explore avenues that are mutually rewarding. Our minds act as multi-agent systems, where neural nets with different perspectives communicate in reaching group decisions through arbitration. Our minds also combine selections from different perspectives, which can lead to innovation and new ideas. The same cognitive principles worked slowly in evolution to select the pathways of bacterial chemotaxis. This same set of rules that works in our brains can be considered in an evolutionary scenario as a long heritable learning process to seek a better future. Sharing values, goals and motivation is what should happen when falling in love, which is done to secure a better future. Mentoring, like loving, is a two-way interaction in communication on how process facilitates progress in attaining goals.

Rewards in the future establish the bond with a motivator whether it is a person or a resource in the environment. In our subjective world, is a mentor the experience creating a sense of wonder that motivates the path to knowledge? A living system in a complex dynamic environment faces the same challenge as a student in a university who finds excitement through a professor's teachings. Universities, as an environment, present students with motivation opportunities to explore. Progress towards goals provides values as knowledge. The knowledge from a variety of goals also gains relevance with other interests and the opportunity and ability to interrelate these with your life. Does environmental variability considered as opportunities present the same general plot to 3.5 billion years of evolution?

However, even though all life must find adaptive rewards in its learning, we humans have the ability to tell stories about the process. We have been provided with unique potential crafted by billions of years of evolution and should employ this skill thoughtfully. We can now reflect on how this potential emerged from physics to light the fire of life.

A Natural History of Cognition.

> *"The large and important and very much discussed question is: How can the events in space and time, which take place within the spatial boundary of a living organism, be accounted for by physics and chemistry?"*

> *"The arrangements of the atoms in the most vital parts of an organism and the interplay of these arrangements differ in a fundamental way from all those arrangements of atoms which physicists and chemists have hitherto made the object of their experimental and theoretical research."*

WHAT IS LIFE? Erwin Schrödinger 1944

Science strives for explanations of phenomena that are compatible with what is known in philosophy as ***physicalism***, where all the stuff comes from the Big Bang. As a naturalist I ask questions of nature, and my book is an attempt to resolve the mystery that was posed by Schrödinger with these two statements. He spoke of the essential difference of living systems as life's dependence on maintaining its highly structured order and coined the term *negentropy* to characterize both life's complexity and the need to continually feed on negentropy to maintain itself. He also told us you could tell life because it kept on behaving in this quest. Life is a complex structure of matter and energy that is not predictable from the laws of physics, since it keeps behaving for a purpose. A living organism must behave to occupy a favorable environment where it can access energy and materials to maintain and replicate the complex structures of its body. These key activities are referred to as embodied, situated, and enactive in the cognition of behavior. That simply means that the body of the individual must be in an environment and must interact with it for its own purpose. In biology this is the study of homeostasis and behavioral ecology. For us to currently exist and behave required evolving to do these same things but in ever more complicated ways. Life and evolution are entwined in cognitive studies as adaptive behavior. Life began as biochemistry and much later evolved rapidly responsive nervous systems. Making behavior adaptive requires practice and correction of errors in both evolution and learning processes. Adaptation takes place at evolutionary, epigenetic, and neural levels, which, although they occur over different time frames, nonetheless can be treated by the same general theory.

The external environment is complex and variable, and, to succeed in its environment, the organism must selectively observe events and predict their causes and effects. This predictive ability is acquired through use of processes that function according to Bayesian inference and employ models that recursively interact with the environment through perception and actions. The story begins with evolution establishing an informational link with a beneficial effect, which permits intervention on a cause–effect relationship for a resource. The model must perceive information from resources in the environment that link to a knowledge model within the organism that interacts with the resource in a purposefully favorable way. The models are adaptively flexible, and organisms adjust them to fit the variability encountered in resource availability. Their functions parallel generative models in artificial intelligence. This perspective on models provides more understanding in the creation of a fusion of the hardware of physics with the software of evolution for programming behavior. This view of behavior enables a comparative science of cognition that works from biochemical evolution to robotics and provides a broad view of cognitive function on a wide variety of behavioral platforms.

From the perspective of my book, *Life tells physics how to behave, and the physics tells life how to feel*. We can learn many lessons from this evolutionary story, and this critically applies to how to improve our behavior.

> "Education is not preparation for life; education is life itself."
> John Dewey

Perhaps these are prophetic words. Is all life based on learning better behavior? Life is the use of prior experience to figure out how to behave adaptively. Before life, stuff just happened; then, some of the stuff began doing things for a purpose. Goal-directed behavior, **teleology** or **teleonomy**, appeared with biology and has kept it going for nearly four billion years. Life began agency and the correction of errors for improved likelihood of obtaining goals. Information used as evidence, knowledge, and purpose triggered Earth's Enlightenment. In life an individual's knowledge adapted the authority of physics for its own use. What is a necessary set of conditions for an intervention on the ongoing physics of the world in order to employ it for an adaptive purpose? How can it happen with the stuff that had, for the billions of years before life arose, just followed the rules on what should happen? Just as an electron can simultaneously be a particle and a wave, it is also possible for molecules in a living system, which provide its structure, to be used as the physical carriers of the information to shape a behavioral model.

The deep history of cognition.

The behavior we associate with the functions of living systems emerges as a complementary aspect of the elaborately structured chemistry that builds these systems. Evolution structured molecules into complex assemblages that have the needed agency to acquire energy and material for replication. To succeed in these exploitations of environmental resources, the properties of molecular assemblages and their interactions in space and time came under the control of selected models. The flow of information between environment and organism enabled development of behavioral models that fostered adaptive performance by living systems. Emergent cognitive functions arose in these generative models. Thus, for these models to fulfill their purpose in living systems requires observation, control, and error correction. Even at a primitive level, therefore, we can regard life as having cognitive abilities. These abilities are associated with evolving the ability to manipulate the ongoing physics for a purpose. Fundamentally, a learning process begins by selecting the right information for writing a program for behavior. Living systems began to create these cognitive processes when biochemical pathways first arose. For example, ATP-generating pathways must "read" the chemistry of the cellular environment to "know" whether to continue breaking down substrates for ATP production. At a higher level of organization, a cell must "read" the chemistry of its surroundings to "decide" on its opportunities for attaining materials needed for energy generation and growth. We therefore can conclude that the structure of these cognitive models for generating agency with adaptive behavior is shaped over a broad spectrum of time and levels of biological organization through observation on success of performance. Correcting errors to optimize performance enables the adaptive use of physics to obtain a required resource. Nervous systems had to wait 3.5 billion years for their turn to contribute to the elaboration of purposeful performance that arose with the original biochemical pathways of cells.

Behavior versus DNA: Who runs the process of evolution?

Common parlance commonly develops aphorisms for use as general explanations of complexity. After Darwin, life became "survival of the fittest." Genetics gave us new adaptive characters as mutations of genes, and life became survival of the luckiest, since mutations were chance events. The structure of genes as DNA and the science of molecular biology placed the information in DNA in control of survival. However, as argued throughout this book, I feel life is better understood as behavior rather than DNA—and life is best viewed as survival of the wisest. The prominence of the role played by DNA and genes has become a widespread religion, and society worships at its temple: *"Religion: the belief in and worship of a superhuman controlling power, especially a personal God or gods."* Below are some quotes from Richard Dawkins, who popularized the gene importance in evolution.

> *"It is finally time to return to the problem with which we started, to the tension between individual organism and gene as rival candidates for the central role in natural selection...One way of sorting this whole matter out is to use the terms 'replicator' and 'vehicle'. The fundamental units of natural selection, the*

basic things that survive or fail to survive, that form lineages of identical copies with occasional random mutations, are called replicators. DNA molecules are replicators. They generally, for reasons that we shall come to, gang together into large communal survival machines or 'vehicles'." The Selfish Gene 1976

"We are machines built by DNA whose purpose is to make more copies of the same DNA. ... This is exactly what we are for. We are machines for propagating DNA, and the propagation of DNA is a self-sustaining process. It is every living object's sole reason for living." Royal Institution Christmas Lecture, 'The Ultraviolet Garden', 1991

"DNA neither cares nor knows. DNA just is. And we dance to its music." River Out of Eden: A Darwinian View of Life 1995

Put these remarks into the context of the first three billion years of evolution. The genes are not replicators; they are replicated by the vehicle's energy and machinery. It is cells that survive along with both genes and their epigenetic behavioral machinery. *I would say that every living object's sole reason for living is to predict how to behave to keep living.* Life discovered many ways to do this, including the use of DNA as a way to replicate informational molecules.

DNA not only predominates the philosophy of science but has spread to many perspectives in societies, notably the widespread ease of belief that our genes determine many of our personal characters and abilities. Many of our current problems that deal with education, motivation, and visions of a hopeful future are colored by belief in the predominant role of genetics. The societal tragedies of racism, misogyny, xenophobia, and class bias find their roots in the belief in determinism by genetics. These beliefs can all change through better behavior.

As a teacher, I have always felt that the ability of a person to motivate others to practice and correct errors would confer to these individuals an advantageous position for improving behavior. This error correction practice is also a positive feedback system since the more you learn the faster you are able to learn, and your improvement is both motivation and reward. This is a fundamental characteristic of our cognitive ability that was discovered by the early biochemical pathways of life as a means of shaping better and better behavior through evolution. In contrast to Dawkins' statement on DNA not caring, I have been trying to make the point that the cell both knows and cares about how it dances to the music that it both creates and monitors for necessary improvisation. DNA doesn't do anything; it lies around waiting to be told what to do by the epigenetic programs built from molecules it provided the information to construct. Survival while you dance requires the flow of information between the components that do the dancing, in order to correct errors and adapt to change by requesting adjustments in the music played by the genes. Prize-winning dancing requires gene regulation by communication between pathways that network the agents involved in the dance.

DNA and genes comprise a library with blueprints for molecules that are directed to be parts of the epigenetic machinery composing the cell, the machinery that really does things. The DNA is told to propagate by the cell, which has decided to propagate itself. Problems arise when the genes don't listen to their normal instructions and participate in selfish decisions, as in cancer cells. The cell is coordinated biochemical behavior, wherein which lies the mystery presented by Schrodinger; "*The arrangements of the atoms in the most vital parts of an organism and* the *interplay of these arrangements differ in a fundamental way from all those arrangements of atoms which physicists and chemists have hitherto made the object of their experimental and theoretical research.*" Behavioral models are these specific arrangements of atoms and molecules that carry information to be used as messages. These instruct the pathways in what to be, what to say, where to be, what to do, and how and when to do it. In cells this all depends on the selections that are chosen to be read from the genetic library, when they are read, and the pattern of reading them. This is epigenetics and molecular biology producing the physiology of the cell.

This is my reason for treating life as depending on the knowledge acquired in building and, then, using the library in the behaviors that take place, rather than regarding the library as controlling the show. The ancestral cell was a replicating vehicle that moved through the world, adaptively sensing and creating behavioral models fitted for the world it interacted with. Long past and recent histories determine the machinery and priorities used now in shaping the behavioral response to maintain homeostasis. The cell is the basis of life and must have a capacity to correct errors in its machinery that programs behavior at the levels of genetics and epigenetics. The cell can exist as an individual, act in social groups or form multicellular individuals. The cell's programs are adaptively variable in accessing the library in the same cell at different times. In multicellularity, different cells with the same genes employ them for different purposes. This is the adaptive use of libraries; diverse interests can access it for different purposes. Genes compose a massive and very complex library. Cells access this library to adaptively maintain homeostasis and respond appropriately to input from their environment. Just for the genome of each of our cells there are six billion of the four bases that write the code. This information can be read and interpreted in many ways for many purposes. The genes are a given and an organism's success depends on programs of epigenetics and how good it is at using the library to facilitate adaptive behavior. This point stresses the importance of building an adaptively useful library, correcting errors when they occur, and using the library effectively in a complex and changing world.

Some concluding thoughts: an overview of evolution through learning.

This perspective on the basis of life and its success over 3.8 billion years stresses that cognitive functions in the earliest forms of life followed the same general rules as our brains currently follow, to work to program our behavior. This essay was introduced as a natural history of cognition and cognition of our biochemistry is how this fits an evolving story of life's uniqueness. This analysis of cognition and its key role in life began when I was informed that I was not consciously in immediate control of my life. I was amazed to discover both decisions and perceptions were delayed by a half second before becoming conscious. My conscious could quickly support or change my behavior in its progress through interactions with the controlling pathways. Even worse to my self-ego was the miniscule informational content of my conscious compared to the enormity of processing that determined it for my review. My consciousness could handle about 40 bits per second, but my unconscious self only worked at hundreds of millions per second.

To find adaptive meaning for the story of my life, I decided that consciousness must act as the critic and editor of the unconscious processes that underlie the conscious state. Consciousness is like the director of a movie who surveys small chunks of the ongoing story for review. The results would be a function of the recursive interaction of how what we created influenced direction and motivation of what we would do next—living our lives was the unfolding of this process. The story we compose as we live our lives is written from chosen episodes that establish our autobiographical memory. As we move into our future, the episodes are subject to editing, enhancing, and revision, influenced by what we do with their use in new experiences. We can edit with stupidity, bias, intelligence, morality, greed, hope, fear, wonder, ideology, questions and a wide spectrum of motivational paths. These edits will play a role in determination of our future.

The neurological mechanisms available for creating an edited autobiography are extraordinarily complex and continue to be revealed by neuroscientists. For example, brain scans have revealed areas that are activated when study subjects in a brain-scanner let their minds wander; this activity decreased when they consciously concentrated on a task. This activity was labeled as the **default mode network** (**DMN**) that broadly activates as if it were the dark energy of the brain, i.e., an abundant component of the system, but one that is quite "invisible" in most circumstances. What does the brain do unconsciously when released from the conscious control? The function of the default mode intrigued the interest of many scientists who extended the research of the DMN. It was found that selected areas of the DMN were also accessed during conscious processing. Perhaps this enhanced activity during mind wandering was associated with the high level accomplishments attributed to our conscious mind. Over a dozen functions have been associated with activity of the DMN. Among these was the management and access to the autobiographical memory. The DMN also triggered attention, thus directing the conscious to interact with meaningful input, which could later be processed to participate in our autobiographical memory. If caused to think, the conscious brain can play the role of the teacher who asks questions to direct learning pathways. The product is fed back into the pathways of unconscious Bayesian processing of the DMN. New interests can later play a role in selection of attention and influence the directions taken by the mind.

The work on the Bayesian brain described a general theory of cognition that can function to predict the best response to input through the optimization of generative models. An agent is defined in terms of its behavior, which is the attribute that distinguishes life. For computer programs, an agent is software that autonomously performs tasks for goals of its host. Multi-agent systems are an appropriate means for modeling complex adaptively responsive systems. The globally connected multi-modal network of the DMN that writes the story we use to run our lives is the ultimate of socially networked multi-agent systems. Our consciousness plays the role of directing the quality of the story. Behaviors have individual goals and shared consequences, thus demanding open communication for optimality. This embeds values, emotions, empathy, rationality, and virtues in behavior at all levels. While we can view this in the light of our consciousness, it is also crucial to build our values into adaptive operators of the underlying unconscious processing that does it all.

Our functionality at getting it right depends on the work of the conscious selection of evidence for processing in the unconscious. The DMN must be supplied with the evidence that can focus attention in adaptive ways. To flourish, behavioral models must monitor and correct their predictions in a complex and dynamic environment. Multi-agent interaction, to be adaptive for all, required compromise and produced communication and arbitration between agents' models. Equating learning and evolution in the framework of the Bayesian principle acting on different time scales provides a general theory for biology and the information systems that have been produced by its actions. This is how we emerged from physics and acquired our unique abilities. Learning by Bayesian inference produces both individual perspectives and shared objective views of how the world works. There is a real world that supplies the evidence for hypotheses on beliefs about how it works related to our behavior. To share valid objective views for important knowledge of the world, we must use evidence from reality. Getting it right is important. After thirteen years of reading and thought, I have brought together a story from evolution, biology, psychology, physics, philosophy, and the emerging field of artificial intelligence to support this position. This adaptive story of behavior with survival value has a general framework that has been formalized by Karl Friston as the **free energy principle**. It is based on Bayesian inference that uses all available evidence to predict adaptive response to challenges and opportunities.

Lineages of life provide stories of success and failure which can serve as lessons that have heuristic value if our species is to have a future. We have come so far from simple beginnings and now have created the ability to look into the future. We can examine the consequences of our actions to understand, predict, and act in order to provide equitable futures for life on our planet. I find hope in the words of Albert Einstein in his reflection on the role of science for humanity, where he emphasizes the need for justice in goals. The success of science in perfecting means suggests that the use of all the evidence for beliefs concerning goals would be rewarding.

> *"Perfection of means and confusion of goals seem–in my opinion–to characterize our age. If we desire sincerely and*

passionately for the safety, the welfare, and the free development of the talents of all men, we shall not be in want of the means to approach such a state. Even if only a small part of mankind strives for such goals, their superiority will prove itself in the long run." Einstein 1941

Future Reading and References.

Kauffman, S. (2008). *Reinventing the Sacred: A New View of Science, Reason, and Religion.* New York: Basic Books.

Mooney, C. (2006). *The Republican War on Science,* New York: Basic Books.

Nørretranders, T. (1991). *The User Illusion: Cutting Consciousness Down to Size.* New York: Penguin Press Science.

Wilson, E.O. (1998). *Consilience: The Unity of Knowledge.* New York: Alfred Knopf.

GLOSSARY

Abduction: Inference from experience as to the simplest and most likely hypothesis to explain a problem.

Access consciousness: Ned Block considered two types of consciousness. Access consciousness is used in reasoning and for direct conscious control of action and speech. While phenomenal consciousness results from sensory experiences. We combine these as part of an agent's behavior.

Active inference: The recursive set of actions and observations in selecting evidence to optimize and control a model in its interactions with input. Either the action or the model or both may be adjusted by correcting errors in performance

Adjacent possible: Potential ways the present can be employed and shaped for new purposes as innovations in structures or behaviors.

Affordance: Entities in the environment as percepts of their potential for use in behavior or signals to respond to in adaptive ways.

Agent: An agent is an autonomous system that acts to achieve its goals.

Artificial Intelligence (AI): The discipline of machine learning and problem solving that mimics cognitive functions associated with the functioning of human minds.

ATP (adenosine triphosphate): The organic molecule that provides the energy to drive many processes in the living cell.

Autonomous: Systems that act to determine their own perceptions and functions.

Autopoietic: A system capable of reproducing and maintaining itself. The noun describing such a process is **autopoiesis.**

bacteria: In lower case this is the common term for the first form of cellular life, which lacks a nucleus and whose DNA is a single circular chromosome. They early split into two Kingdoms: Bacteria and Archaea.

Bayesian: Refers to statistical methods based on Bayes' theorem, which are used to update the probability for a hypothesis as more evidence becomes available.

BDI: Acronym for **belief-desire-intention.** Developed by Michael Bratman as a model of human practical reasoning as a way to account for decisions and actions.

Behavior: The way in which a system acts in response to a particular situation or stimulus. The act of an agent doing something for its own purpose.

Belief: Predictability that an agent has the knowledge and ability to perform a behavior.

Boundary conditions: For unique mathematical analysis, constraints or input scenarios are established as boundary conditions for the system's study.

Causality: The case where one event contributes to occurrence of another event.

Cognition: The process of acquiring knowledge and understanding through the senses for the purpose of interaction with another system.

Consciousness: The process of an agent to observe, monitor, and interact with its ongoing behavior of some level of complexity. An agent's observation of a challenge or opportunity and the use of active inference to create and/or process information in its resolution.

Counterfactual: This expresses alternative causal chains for outcomes of actions that have happened or could happen and is used for analyzing causation, correcting errors, and planning for the future.

Decision: The selection and energizing of the set of behavioral models for a course of action.

Deep Neural Networks (DNN): AI machine deep learning methods based on hierarchical stacks of artificial neural networks for representation learning.

Default mode network (DMN): A multimodal network of interacting brain regions that becomes more active when a person is not focused on the outside world.

Dualism: Mind–body dualism is the view in the philosophy of mind that mental phenomena are non-physical, or that the mind and body are distinct and separable.

Dunning/Kruger effect: the common situation in which the less a person knows about a subject, the higher the level of confidence he/she has that they are correct.

Effective Field Theory: Our Universe exists at hierarchical levels of dimensions, energy, and interactive grouping of constituents. Higher levels of dimensionality often have emergent properties not predictable from the lower dimensional level. We observe emergent properties at higher dimensions and use these to construct effective scientific theories at that level. **Renormalization group** is a mathematical justification of the changes of a physical system when viewed at different scales. Classical physics is an emergent effective theory of quantum mechanics.

Embodied, Embedded, Enactive: The cognitive approach that refers to the concept that perceptions, beliefs, and actions derive from direct experience of the senses and the body interacting with its environment. The body and sensed environment are extended parts of the mind. This fits our concept of the generative model using active inference to minimize the free energy for purposeful behavior, where the generative model is the individual adapting to its niche by evolution and learning.

Emergence: As in effective theory, this a property or phenomenon that occurs at macroscopic scale but not at its underlying microscopic scale in reductive analysis.

Entropy: A system's lack of order or predictability or ability to do work; 2nd Law states there is a gradual decline into disorder. Living systems are highly ordered (Schrodinger's negentropy) and require input of negentropy for maintenance.

Epigenetics: A cell's orchestrated chemical reactions are regulated by their roles in turning on and off parts of the genome to meet demands and changing conditions. Sets of these regulatory pathways are heritable with the cell's division as well as the genome.

Eukaryotes: The symbiotic fusion of a cells from the Bacteria and Archaea domains of life produced the diversity of life composed of cells with a nucleus, chromosomes, mitochondria, and plastids, which comprise the diversity of protists, fungi, plants, and animals.

Evolution: Evolution is change in the heritable characteristics of biological populations over successive generations through selection of adaptive performance.

Executive function: Working memory's executive function evokes a set of processes that have to do with high level planning and problem solving. This involves marshaling the necessary resources and motivation in order to achieve a goal as well appropriate interactive behavior to reach the desired result.

Explanation: An attempt to create a causal chain of knowledge using evidence to justify a hypothesis to provide understanding of how and why for something.

Free energy: This is the mathematically tractable stand-in for model performance, which approximates model evidence and must be minimized to improve performance.

Free will: the concept that our capacities as an agent include the ability to consciously choose the action we will take.

Generative model: Predictive belief models that train themselves through active inference to select evidence from a complex dynamic environment to learn the most salient information to generate representations for behavioral interactions.

Hard problem: David Chalmers posed this as the mystery of explanations for the qualities of our subjective experiences. Thomas Nagel posed this as "What is it like to be a bat?" This is the central theme of my book, which I treat as the general problem of what it is like to be alive and an adaptively behaving system so that consciousness is an emerged property of life.

Heritable: In the broad sense this refers to the ability to pass traits or their elements from one system to another.

Holism: Many systems are composed of many parts with different properties and histories that interact in ways that are not well explained by reductive analysis. These systems must be examined from different perspectives and understanding approached by synthesis.

Homeostasis: The ability to maintain a relatively stable internal state that supports the system's requirements despite changes in the environment.

Hypothesis: A proposed solution (abduction) to a problem based on the evidence at hand which is the prior in Bayesian analysis.

Idealism: A philosophy that "reality" is based on human understanding and/or perception and our reality is a mental construct.

Implicit learning: the adaptive shaping of behavior and understanding, without awareness and reportability of what is learned; thus, it is intuitive in nature.

Inference: The process of selecting evidence to construct or optimize the probability of a hypothesis or prior.

Information: Data that have been given meaning relating to its purpose in a knowledge system and is represented by a symbol that can participate in a message.

Information bottleneck: This involves the network shedding input data of extraneous details by squeezing it through a bottleneck and selecting those most relevant to the concept.

Intention: the decision that an agent will perform a behavior.

Intervention: An action taken to change what would have happened in a system.

Knowledge: An interactive model that is built from information and can either use other information for a purpose or convert data to information for its own use.

Krebs cycle: The sequence of biochemical reactions by which most living cells use substrates to generate energy in the form of ATP.

Libet delay phenomenon: the ~one-half second delay between our feeling aware of a perception or of making a decision and the subconscious neural processes that, in fact, create the particular conscious state we feel.

Life: The continuous set of systems emerged from the physical world that autonomously developed behavior to use energy and resources to replicate and sustain their processes. This is treated as evolving biochemical agency competing for resources as it diversifies and becomes ever more complex.

Manifest image: Wilfrid Sellars proposes this as the framework in terms of which we ordinarily observe and explain our world as persons and things before the intrusion of the esoteric scientific image.

Markov blanket: This partitions a system into internal states and external states, where the sensing and action of active inference pass through the blanket. Life has produced hierarchical sets of Markov blankets self-assembled and integrated into composites with their own Markov blanket

Meaning: In cognition and semantics the symbol grounding problem relates to how words or symbols as referents get their meanings. In adaptive behavioral systems the meaning comes from the transformation of data to information for a purpose.

Mind-body problem: This is a long-standing philosophical issue that, in essence, inquires how 'inanimate matter' can generate the mental phenomena we experience. Some putative "solutions" to the problem, such as the one proposed by Descartes in the 17th century, propose separate "physical" and "mental" materials in the Universe.

Metabolism: The set of biochemical reactions that occur in an organism in order to maintain its life.

Model: A representation of some properties of an entity constructed for an interactive purpose.

Normative: This refers to statements that make assertions about values and how things should be rather than factual statements about reality and are often used to judge counterfactuals or to plan for the future.

Observation: A system that transduces energy from another to create symbols as information that represents properties of the observed system.

Ontology: The discipline that deals with reality, what exists, properties, origins, relationships, categories, and their definitions as concepts. In science it becomes definitions of concepts and relationships that exist for an agent or group of agents.

Perception: Representations of our sensory modalities from the processing of observations into behavioral models.

Phenomenal consciousness: Ned Block treated two types of consciousness: phenomenal consciousness results from the representations and the feelings of experience. Access consciousness is above sensory experiences.

Physicalism: The philosophical position that everything that exists is based on the products of the Big Bang and the particles and forces of the Standard Model.

Poetic naturalism: Our world is very complex and consists of many hierarchical levels. Sean Carroll proposed there are a variety of ways to talk about the world using language dependent upon the aspect of reality being discussed.

Posteriors: In Bayes theorem this is the adjusted probability of a belief based on the use of new evidence.

Precautionary Principle: In essence, this principle says, "Be aware of how negative an outcome of some action might be, and design means for preventing this worst case scenario (or simply an awful scenario) from occurring." Reducing greenhouse gas emissions is an obvious—and important—instance of this Principle.

Priors: In Bayes theorem, this is the probability you have assigned to a belief.

Probabilities: Numerical statement on the likelihood of an event or of a prediction happening and also applies to confidence level in a belief.

Quale: The experience of qualities or feelings, like redness or pain, as considered independently of their effects on behavior.

Recursive: The repetitive application of a procedure which is important in adaptation of behaviors in providing feedback and feed forward to optimize performance.

Reduction: For science and explanations in general this is the attempt to provide understanding through a reductive downward causal chain of more basic elements already known.

Renormalization group: Mathematical justification of the changes of a physical system when viewed at different scales. A coarse-graining scheme that allows for the extraction of relevant features as a physical system composed of aggregates of microelements is examined at different length scales. As the scale increases it creates higher levels with emergent properties requiring their own theories.

Representation: Representation is the use of signs that stand in for and take the place of something else. Observation, measurements, perceptions, models, consciousness, beliefs, priors, hypotheses, and metaphors all derive from the concept of an observer as a process of creating a cognitive informational symbol with meaning that represents reality external to the system.

Salience: This is the state of an event or observation being important, of value, noticeable, attentional, or surprising.

Semiotics: The study of signs or symbols as having meaning for their user in representing a property, object, idea, or relationship in a communication system.

Supervenience: In philosophy, supervenience refers to a relation between sets of properties or sets of facts. X is said to supervene on Y if and only if some difference in Y is necessary for any difference in X to be possible.

Surprisal, Surprise, Self-information: When something happens we don't expect, our surprise directs attention and may call for response. This is fundamental to behavior and evolution. Bayesian surprise measures the differences between posterior and prior beliefs of the observers. This relates to sensory processing, adaptation, learning, conscious processing, and decision-making.

Symbol Ground Problem: how words and symbols gain meaning which relates to their meaning and reality.

Symbolic systems: Systems which generate symbols as information to represent meaning in the solution of problems. Symbols are used in algorithms for computation and sharing information within an agent or group of interactive agents.

Symbolic Systems Approach explores the relationship between natural and artificial systems that represent, process, and act on information.

System Awareness: A system is self-aware if it is continuously perceiving and generating meaning from the models it continuously updates.

Teleology/Teleonomy: We have treated life as behavior based on the requirement of using some part of its environment as a source of energy and materials for its survival. This is based on generative modeling that creates internal models of the system that the organism interacts with. This implies that the functions of the processes of the system are directed to purposes or goals; in the language of poetic naturalism, teleology emerged with life. Those who object to the concept of purpose in nature describe this goal-directed behavior as **teleonomy**, which doesn't carry the philosophical baggage that the term teleology has acquired over the more than 2,000 years of its use by philosophers. Thus, teleology is defined as a reason or explanation for something as a function of its end, purpose, or goal. Teleonomy, in contrast, is the quality of apparent purposefulness and of goal-directedness and the book is symbol grounding for purpose in behavior.

Theory of Mind: Folk psychology as the ability to attribute mental states such as beliefs, desires, emotions, and intentions to oneself and to others.

Type 1 error: A type 1 error is also known as a false positive and occurs when a researcher incorrectly rejects a true null hypothesis. The probability of making a type I error is represented by your alpha level (α), which is the p-value below which you reject the null hypothesis.

Type 2 error: A type II error is the non-rejection of a false null hypothesis.

Umwelt: The self-centered world of an organism that is its environment as sensed and experienced by the individual.

Unconscious: We have defined consciousness as the assembly and use of behavioral models and unconscious would refer to their use when not chosen to run through working memory and escapes attention and reportability of our autobiographical memory.

User Illusion: This expression refers to an analogy between (i) what is displayed on a computer's screen versus what occurs in the background through activity of the hardware and software and (ii) what occurs in our consciousness versus our subconscious brain activity. Common to both the computer and our brain is the display of a tiny fragment of the massive hidden information processing that creates what we consider the product of the machinery; and, in both cases, what we see is selected by the processes of the machine. It is also of interest that we use what we see on the screen to influence what the computer will do for us in the future. The expression was coined by researchers at Xerox Corporation's PARC facility and made popular by the Danish journalist Tor Nørretranders in a book by this name.

Value: This is the measure of the benefit derived from a behavioral process and thus involved in motivational level to employ and optimize it. This is associated with surprise, reward and utility and also has a relation to the predictability of behavioral performance.

Vitalism: Vitalism is the belief that life differs from non-living entities due to some non-physical element or is organized and controlled by different principles than are inanimate things.

Working memory: Working memory is evoked by attention and involved in goal-directed behaviors. Three subcomponents are proposed: phonological loop, visuospatial sketchpad, and interaction with the central executive function. It also interacts with the default mode network and long-term memory storage and retrieval.

ACKNOWLEDGEMENTS

How does one acknowledge 90 years of contributions to this book? The book is an expression of extracts from my life in a document shared in language. I feel I am like a tree that has been nourished by the resources of an environmentally rich and diverse ever-changing world. Early in my life I began a love affair with the natural world which remained throughout my life. Nature is motivational in my focus of attention. I became enriched by a system of deepening roots that supported branches that grew towards many wavelengths of themes of information. I was drawn to grow in these directions as they reflected attributes of people I admired and wished to know better. I express my appreciation for all these contributions, personal and vicarious, that contribute to my soul. Creative abilities flower and produce fruit when information from many themes is assembled into evidence. I will try to access elements from my soul to express my appreciation to those who helped an explorer choose paths leading to this book.

The organisms and people I admired that enriched my life would fill the pages of a book. For this book I acknowledge a miniscule subset of those who have played roles in making it what it is. Support for reading and communing with nature were parental contributions that began the journey to the book. There are many teachers and mentors but Ted Bullock, who I was sent to as a new biology student at UCLA, indoctrinated me for eight years in reading and knowing all the evidence for your statements. At Stanford Dave Regnery, Don Kennedy and Don Abbott were paradigms of teaching that helped me in communicating with students. Undergrads and grads that I learned for and from were a massive contribution to what I wrote in the book. They number in the thousands but only a few will be named. My tenures with the Monterey Bay Aquarium, MBARI and Sea Studios were massively participatory in my directions of development. As documented in the book the return to teach in Holistic Biology was pivotal for the book's being.

Before I acknowledge the people who assisted in the creation of the book I must recognize that it was Hopkins Marine Station that made it happen. This is Stanford's marine lab on the shores of Monterey Bay, and I relocated there in 1974 from the main campus. The setting facilitates research ideas, teaching, close interaction of students and faculty including social lives. Deep meaningful friendships are established that continue for years and decades. The environmental setting stimulates reflection, diversity, creativity and a synthesis of science and nature. My tenure at Hopkins led me along new paths that enriched and redirected my life. It provided ideas, motivation and people that led to the Monterey Bay Aquarium, MBARI and Sea Studios. The group of people below that played an intimate role in the book's beginnings and completion all share the Hopkins' connection.

The book spent about twelve years in its development as my reading and writing took me along an interconnected maze of paths. It went from primarily environmental messaging to how societal problems related to human behavioral problems and the failure to correct errors in societal performance. Errors proliferated as humans failed in rational virtuous communication as they created problems. I shared parts of the content with friends who cautioned me on difficulties in style and content. As I continued to edit and rewrite and add topics one friend suggested forming a team to facilitate getting the book published. They are a magnificent, diverse, and talented group that enacted a metamorphosis of my text that provided form and beauty and provided wings so it could fly from my computer.

My writing spoke better to me than to a broader audience and that was remedied by three close friends. George Somero became a friend in the early 60s when he began his PhD at Stanford. He interested me in biochemistry, a subject I had avoided, and I followed his work at other institutions. He returned to Hopkins in the 90s as an environmental biochemist also with interests in philosophy and how the brain works. Jim Watanabe, I came to know as an undergraduate and maintained close association as he did graduate work at Hopkins. Jim then went next door to the Aquarium and returned to Hopkins in my position on my retirement. He is an ecologist, excellent naturalist and also teaches statistics and experimental design. I envy his teaching skills. Tierney Thys, I met and worked with at Sea Studios on documentary films and developed appreciation for her skills of communicating science to a broad audience. Ideas are not ideas if they can't be communicated and also ideas are beginnings of hypotheses that always benefit from editing especially by a diverse group of editors. My book was fortunate through their guidance and I am grateful to have them as collaborative friends.

Michael DeLapa is a longtime friend from Hopkins who then went to business school. He emerged and became active in politics. He provided business guidance for Sea Studios and located in the Monterey area as an active social host. Mike recognized that societal environmental action must create a fusion of science, business, politics, advertisement, and communication for effective action in groups with differing agendas. He is an ultimate organizer and I have greatly profited from his environmental approach. Mark Shelley became a friend as an undergraduate in many courses where he was an outstanding student for many reasons other than his grades. That he had greatly profited from his education was revealed in creative innovative expertise in interpreting science and nature in documentaries. He introduced the use of the newly emerging high quality video. After doing the interpretive pieces for the Monterey Bay Aquarium he established Sea Studios that was fundamental to me in what communication involved and what I have learned from him spans fifty years. John Cooper was a student early in my time at Hopkins who also became part of a social group that included me. He was highly participatory in discussions and we have had many. John served as dive master in my summer course in Subtidal Ecology for several years until going to med school and emerging as a doc who continued visits and enriched my life. I still reflect on his challenge to justify humans as separate from nature and this is part of my book's theme. In the chapters on virtue and spiritualism I stress the importance of humans as part of nature; so, you convinced me. It was Coop who sent out the call to organize the group I am lauding.

Nancy Burnett became a close friend well over 50 years ago when she married Robin Burnett, another close friend. Nancy has during that time facilitated aspects of my life that were causal in the book happening. A partial list of my experiences she played a deterministic role in includes Hopkins, the Aquarium, MBARI, Sea Studios and the cruise to the Sea of Cortez resulting in Holistic Biology. There is much more that relate to teaching, science, nature, environmental issues. She loves nature and I was deeply impressed by her devotion to fungi documented by wonderful images and knowledge of their biology. Nancy has spent her life involved in meaningful decisions about how to improve the world. As a mother and grandmother, she has deep appreciation for intergenerational justice and what the future should look like. I along with many others deeply appreciate her friendship.

Greg Baxter is my son and I have known him the longest of the group members. His acknowledgement for the book derives from overlapping interests in biochemistry, xenobiotics, and environmental and health issues. His approaches have involved roles of science, education, innovation, entrepreneurial approaches, and the importance of group process. We have shared many discussions over libations and the importance of the diversity of perspectives necessary in solving complex problems with diverse causes and consequences involved in solutions.

Susan Harris has been my bonded partner for the most rewarding part of my life which began when I turned fifty. I have found it revealing that all the best things I have done and participated in occurred during this period. There must be an explanation and the book has helped me understand. You do good things when you have empathic motivation. I treat this in my chapter on spiritualism where a sense of wonder invokes a desire for deeper understanding and participation. Susan creates an overflowing sense of wonder that facilitates my other behavior. Also, in Chapter 10 Susan can be seen as my mentor in opening my mind in many new directions where exploration can lead to discovery. I thank you and treasure you Susan as we explore and appreciate the world.

This is an important but tiny subset of all those who have participated as my friends in enriching my soul and contributing to my perspective of life. One of these that could have greatly helped in this book left us to soon. Rafe Sagarin infused my soul with many elements that shaped my emerging beliefs and his personal editorial role would have added beauty and truth. I thank all of my friends for their gifts to me for a full rich participation in life.

ABOUT THE COVER

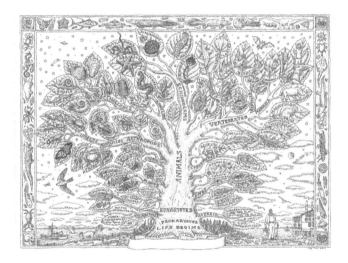

The Tree of Life is by artist Ray Troll. In his studio on a hill above Tongass Narrows in rainy Ketchikan, Alaska, Ray Troll creates fishy images that swim into museums, books and magazines, and onto t-shirts worn around the world. He draws his inspiration from extensive field work and the latest scientific discoveries, bringing a street-smart sensibility to the worlds of ichthyology and paleontology.

The Oldest Tree
By Chuck Baxter

The Tree of Life is also the tree of knowledge. Knowledge is the ability to use information to solve a problem. Life was faced with the problem of maintaining its existence by acquiring energy and resources from its environment. Life creates information to support adaptive behavior. Knowledge models are memories of information as symbols employed to represent relevant properties of the interactive environment. Models are built from selected percepts of the behavioral target for recognition and predicted adaptive action. Symbols are used in programming behavior and communication within and between models. Their ultimate level is the human mind. Life evolved as a symbolic system to interpret causalities in the perceptual world. Life constructs models to represent the structure and properties of the experienced world. Our language has enabled sharing our perceived and learned view of how our world operates. We assume we experience properties of a real objective world but our experience is limited to participation in a subjective world that we create as observers. Life interactively behaves in their adaptive niches selected to maintain support for their life processes. The Tree of Life germinated and sprouted with the life's origins nearly 4 billion years ago. Through an amazing proliferation and diversification from an ancestral cell evolution has produced the biological world including humans who now dominate the landscape. Each group ascended and grew the Tree by learning to behave and correct errors to behave better as conditions changed. The Tree grew and branched by lineages of species adapting their behavior to solve problems and predict how to survive in a complex dynamic environment. Each branch, twig and new bud creates their individual subjective interactive world selecting favorable paths and correcting course on detecting errors in behavior. Each life has an observer-dependent cognitive story of how to respond to input with their behavior. In our species and current situation these stories are diverse and variable in evidence used even in similar problem areas. Our created worlds arise through our life experience. We share experience with language, communication and coordination of social interactions that are now processed by our variety of media. Those who try to understand how there could be different representations of the real world by analysis of evidence used will fail. Humans select much of their evidence subjectively biased by their previous beliefs. Our brain's representation is not of a physically real world so much as it is a meaningful world for interaction according to beliefs and desires. Arguments about the real world and what evidence plays a role in its interpretation evaporate when the problems of a group are considered as Bayesian inferential generative models. Getting it right, equity and justice for the whole group and for future generations requires agreement on the relevance of evidence. Bayesian analyses using all the evidence and their realistic probabilities provide the most objective beliefs for action then followed by adaptive management. When the behavior choices apply to humans as a species and a future for our lineage is the desired goal evolution has provided us with the tools. We must envision an open and hopeful future as a social unit or continue on the destructive path of tribal warfare. The Tree of Life has flourished with unbounded creativity and produced the creativity of mind and culture as the ultimate wonder. We should use our legacy wisely and justly.

CPSIA information can be obtained
at www.ICGtesting.com
Printed in the USA
LVHW040932101120
671142LV00002B/79